Teche

BOOKS BY SHANE K. BERNARD

The Cajuns: Americanization of a People

Cajuns and Their Acadian Ancestors: A Young Reader's History
(available in English and French editions)

Swamp Pop: Cajun and Creole Rhythm and Blues

Tabasco: An Illustrated History

AMERICA'S THIRD COAST

Carl A. Brasseaux and Donald W. Davis, series editors

Teche

A History of Louisiana's Most Famous Bayou

Shane K. Bernard

University Press of Mississippi / Jackson

This contribution has been supported with funding provided by the Louisiana Sea Grant College Program (LSG) under NOAA Award # NA14OAR4170099. Additional support is from the Louisiana Sea Grant Foundation. The funding support of LSG and NOAA is gratefully acknowledged, along with the matching support by LSU. Logo created by Louisiana Sea Grant College Program.

www.upress.state.ms.us

The University Press of Mississippi is a member of the Association of American University Presses.

First printing 2016

∞

Library of Congress Cataloging-in-Publication Data

Names: Bernard, Shane K., author.
Title: Teche : a history of Louisiana's most famous bayou / Shane K. Bernard.
Description: Jackson : University Press of Mississippi, 2016. | Series: America's Third Coast series | Includes bibliographical references and index.
Identifiers: LCCN 2016005806 (print) | LCCN 2016024272 (ebook) | ISBN 9781496809414 (hardback) | ISBN 9781496809421 (ebook)
Subjects: LCSH: Teche, Bayou (La.)—History. | Teche, Bayou (La.)—Environmental conditions. | Bernard, Shane K.—Travel—Louisiana—Teche, Bayou. | BISAC: HISTORY / United States / State & Local / South (AL, AR, FL, GA, KY, LA, MS, NC, SC, TN, VA, WV). | TRAVEL / Essays & Travelogues.
Classification: LCC F377.T4 B47 2016 (print) | LCC F377.T4 (ebook) | DDC 976.3/3—dc23
LC record available at https://lccn.loc.gov/2016005806

British Library Cataloging-in-Publication Data available

To Amy Lançon Bernard,
born and reared along the Teche;
in memory of Glenn R. Conrad,
historian of the Teche Country

There exists no known river on the globe with traits of exact analogy to the Teche; many of its features are peculiar to itself. . . . [A]nd for that simple reason, it is almost impossible to describe the Teche, in language conveying clear conceptions of the object; as there is no river with which it can be correctly compared.

—**William Darby,** *The Emigrant's Guide to the Western and Southwestern States and Territories* (1818)

Contents

Bayou Teche flows for 125 miles through the south Louisiana regions known formerly as the Poste des Opelousas and the Poste des Attakapas.

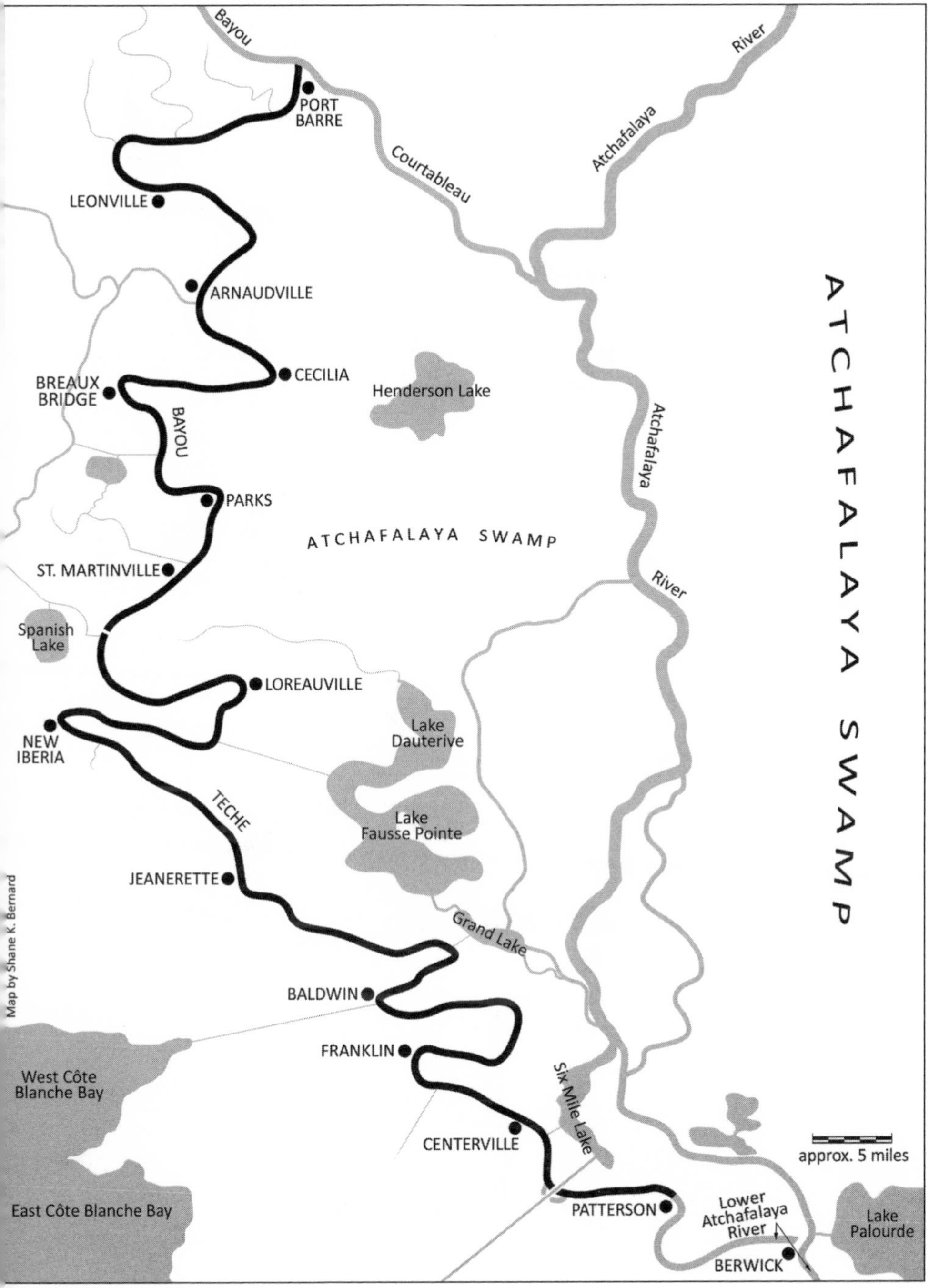

The entire length of Bayou Teche from Port Barre in the north to Patterson in the south.

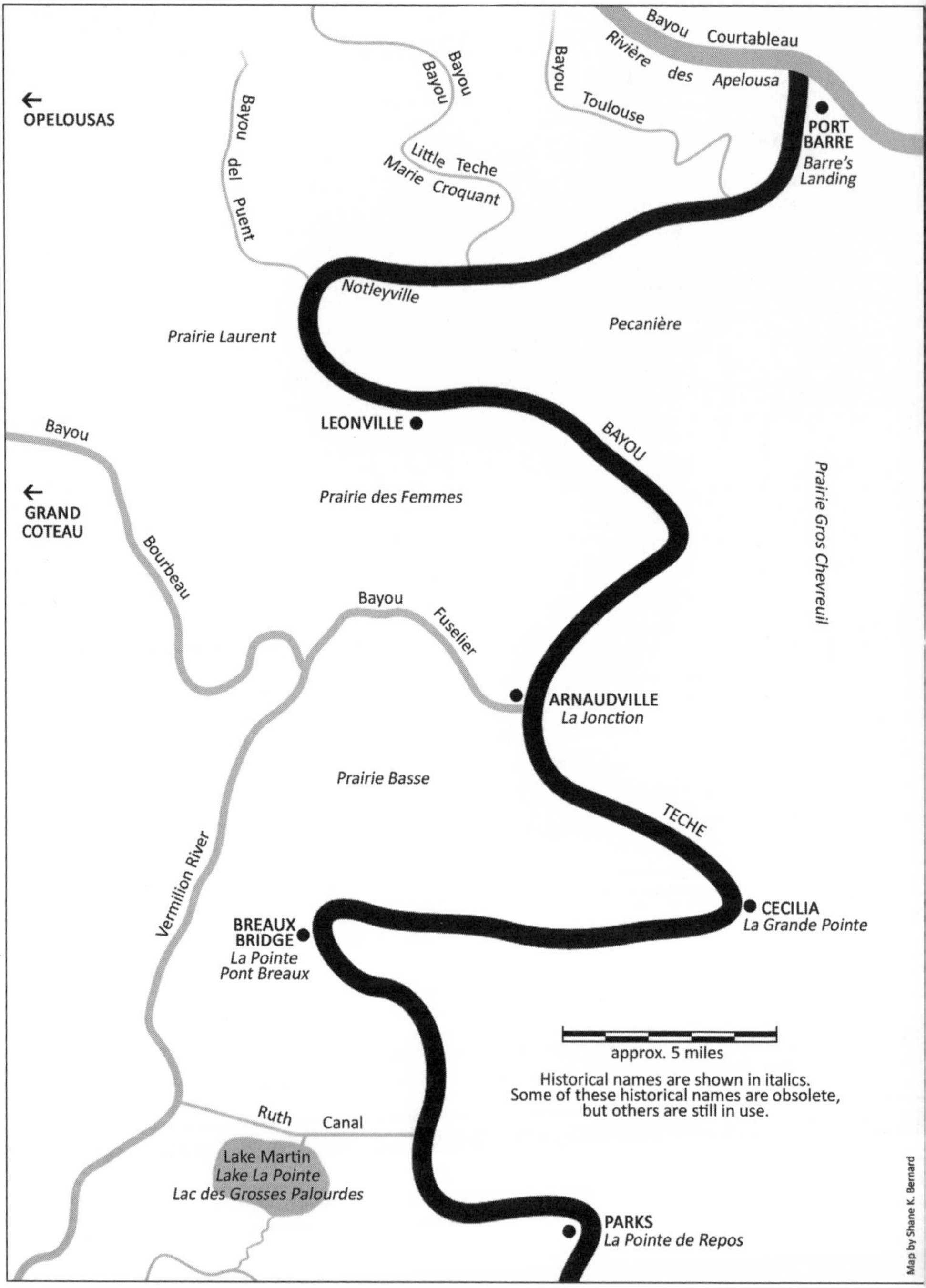

The upper Teche from Port Barre downstream to Parks.

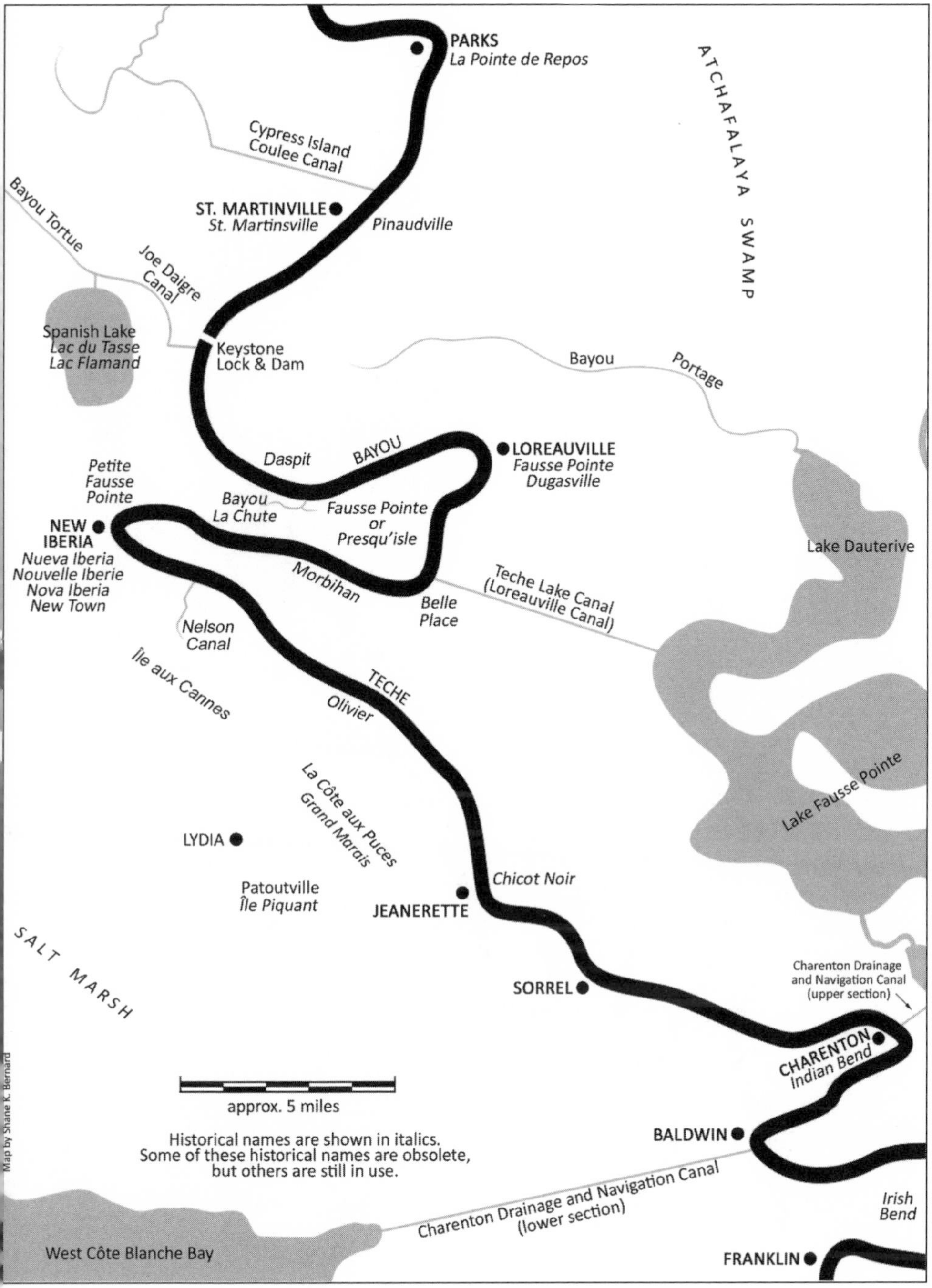

The Teche from Parks downstream to Franklin.

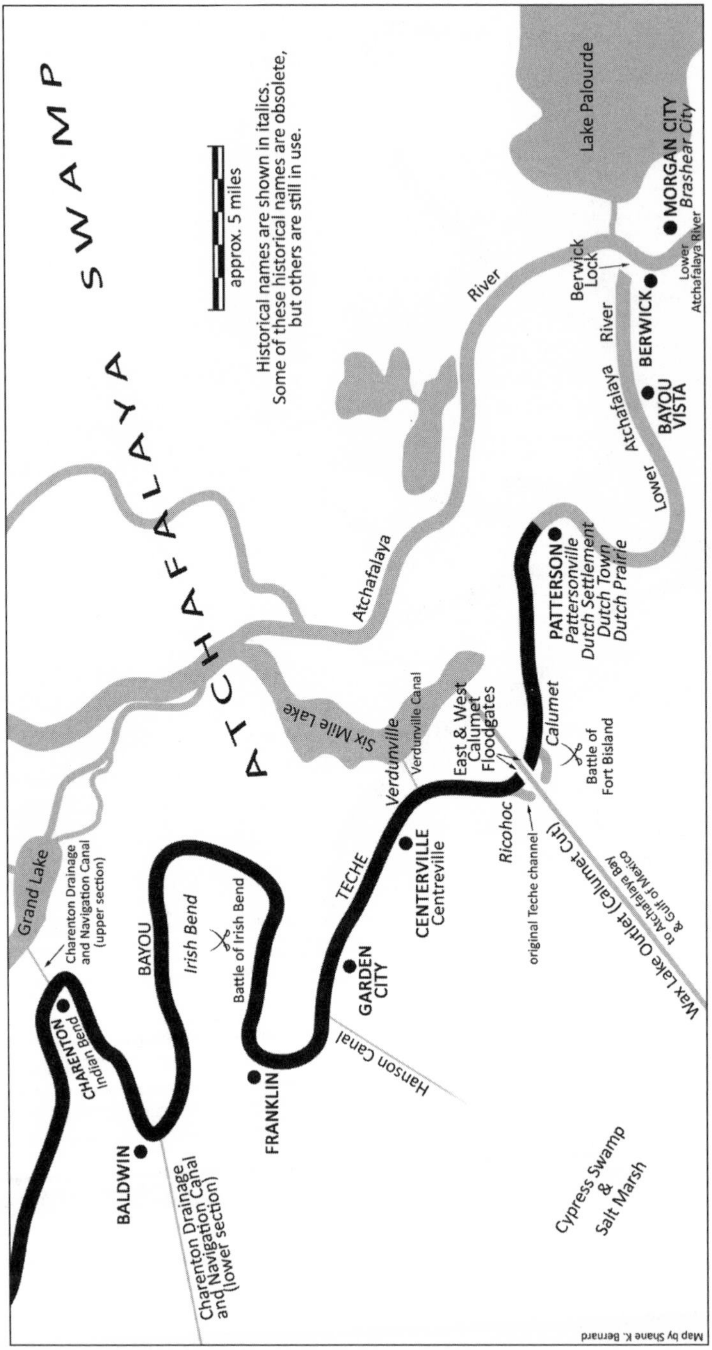

The lower Teche from Charenton downstream to Patterson; below the Teche runs the Lower Atchafalaya River.

Teche

Introduction

A south Louisiana native, I have lived most of my life only a short distance from Bayou Teche. For nearly the past two decades, in fact, I have resided only about three blocks from the bayou. After reading Mark Twain's *Life on the Mississippi* a few years ago, it occurred to me to write a history of the Teche, because it seemed that I had practically in my own backyard a "Mississippi River in miniature" (as I describe the bayou elsewhere in this book). Decidedly *in miniature*: for while the Mississippi at New Orleans discharges about 600,000 cubic feet of water per second, the Teche near Jeanerette discharges only about 415 cubic feet per second. While the Mississippi River runs about 2,300 miles in length, the Teche runs only about 125 miles. And while the Mississippi spans 2,500 feet from bank to bank at the Crescent City, the Teche stretches only about 550 feet at its widest point (just above Patterson). For most of its course, however, the Teche runs much narrower: about 90 feet at Arnaudville, about 145 feet at New Iberia and St. Martinville, and about 225 feet at Franklin.[1]

Despite its relative smallness, the Teche is a historically significant Louisiana waterway—much more significant than its size would at first suggest. William Darby noted this quality in 1817, when in his *Geographical Description of the State of Louisiana* he observed, "This river . . . claims more notice from the political economist and geographer, than either its length or quantity of water would seem to justify." The reasons for this seemingly extravagant attention, according to Darby, were the bayou's fortunate location and agreeable climate, combined with the amazing fertility and yet (at the time) relative cheapness of farmland along its banks. These qualities, he asserted, gave the Teche Country "a decided preference over any other body of land of equal extent, west of the Mississippi." Moreover, the Teche offered itself to explorers, settlers, and travelers as a primitive superhighway leading deep into south Louisiana's often isolated interior—a role the bayou fulfilled until the advent of railroads and automobiles.

But the Teche's significance also stems from something less tangible than geography or economics. For well over two centuries the bayou has intrigued locals and visitors alike, enticing them with a certain mystique, an intangible quality found less distinctly in rivals like the Lafourche, Ouachita, Tensas, Vermilion, Mermentau, and Sabine, among others. Louisiana's first American governor, W. C. C. Claiborne, visited the Teche Country in 1806 and described it to President Thomas Jefferson as "the most beautiful I ever beheld." Another nineteenth-century visitor declared, "One may hunt the world over and never find another bayou Teche; it is a gem dropped in Paradise." "[S]he is a Louisiana grande dame of superb attributes," penned New Orleans author Harnett T. Kane, who in almost haiku tone added, "The Teche smiles and moves in serenity." This allure, partly aesthetic and partly mythic, explains why the Teche has been featured so often in novels, short stories, poems, songs, paintings, photographs, and films.

The bayou, for example, appears in the romantic, idealized paintings of Adrien Persac, Meyer Strauss, William Henry Buck, John La Farge, Charles Giroux, Joseph Rusling Meeker, and Alexander John Drysdale, as well as in the more naturalistic Reconstruction-era sketches of Alfred Waud. In recent decades the Teche has been featured in the works of George Rodrigue and Melissa Bonin (both of whom grew up along the bayou), and in those of Hunt Slonem (who today owns Albania, a stunning antebellum plantation home on the Teche at Jeanerette). The bayou has appeared in films such as Columbia Pictures' 1942 short documentary *Cajuns of the Teche*, whose priceless imagery offsets its at times historically incorrect narration. More recently it figured in the documentaries *Native Waters: A Chitimacha Recollection* (2011), about the Native American tribe living on the bayou's banks at Charenton; *Le Bijou sur le Bayou Teche* (2013), about the survival of Louisiana French along the waterway; and *In the Mind of the Maker* (2016), about wooden boat-building on the Teche.

Likewise, various musical compositions have alluded to the Teche. The earliest known reference dates to 1856, when composer Euphemia

E. Fleurot published sheet music for "Souvenir du Teche Polka," dedicated "to her pupil, Miss Lelia Delahoussaye, Franklin, La." In 1911 famed American composer John Phillip Sousa scored "The Belle of Bayou Teche" with lyricist O. E. Lynne. Written in now offensive "Negro dialect," the lyrics presented a hodgepodge of Old South clichés:

> On de lazy sleepin' bayou,
> On de live oak-shaded bayou,
> Wha' de wateh-spideh spins his silber mesh.
> Wha' de sugah cane am growin',
> Wha' us darkies was a hoein',
> Dere I met my love, de Belle of Bayou Teche.[3]

The bayou has been mentioned most frequently, however, in south Louisiana's accordion-and-fiddle-driven Cajun music tradition. Two distinct Cajun songs, one by Columbus Frugé and the other by Nathan Abshire, bear the title "Valse de Bayou Teche" ("Bayou Teche Waltz"). In Frugé's (1929) the singer rebukes his mistress, "*Si t'aurais voulu m'écouter, chère / toi tu s'rais au Bayou Teche avec ton nég', Chérie!*" (If you would have listened to me, dear / you'd be on Bayou Teche with your lover, dear!). In Abshire's (1970), however, the accordionist pleads, "*Mon beau frère, viens donc m'voir, chèr / viens donc m'voir après mourir au Bayou Teche*" (My brother-in-law, come see me, dear / come see me dying on Bayou Teche). Other Teche-related compositions include Iry LeJeune's "Teche Special" (1950), Austin Pitre's "Bayou Teche Two-Step" (1960), Hadley Castille's "Maudit Bayou Teche" (1989), BeauSoleil's "Le Belle de Bayou Teche" (1994) (unrelated to Sousa's melody), and Pope Huval's "Le Teche" (2010). Many English-language songs also feature the waterway, among them "Bayou Teche" (1969) by Louisiana-born Nashville star Doug Kershaw, "Along the Bayou Teche" (2003) and "Shadows on the Teche" (2011), both by Mark Viator, and "Back to Bayou Teche" (1992) by guitar virtuoso Sonny Landreth (notably covered by the Flying Burrito Brothers). At least three bands have called themselves Bayou Teche or have included the term in their names.[4]

Literature has drawn on Bayou Teche over the years as a setting for novels, short stories, and poems. Henry Wadsworth Longfellow, described as "for more than a century, the most famous poet in the English-speaking world," referred to the Teche in *Evangeline*, his 1847 epic poem about the expulsion of the Cajuns' ancestors from Nova Scotia and their arrival in a new south Louisiana homeland. Generations of schoolchildren memorized Longfellow alongside Byron, Tennyson, and Shakespeare:

> On the banks of the Teche are the towns of St. Maur and St. Martin.
> There the long-wandering bride shall be given again to her
> bridegroom,
> There the long-absent pastor regain his flock and his sheepfold.
> Beautiful is the land, with its prairies and forests of fruit-trees;
> Under the feet a garden of flowers, and the bluest of heavens
> Bending above, and resting its dome on the walls of the forest.
> They who dwell there have named it the Eden of Louisiana.

George Washington Cable referred to the bayou in his works, including *The Creoles of Louisiana* (1885), *Bonaventure: A Prose Pastoral of Acadian Louisiana* (1887), and *Strange True Stories of Louisiana* (1889)—the latter of which included a phony 1822 memoir of a 1795 voyage up the Teche (sold to Cable as authentic by its apparent forger, French Creole author Sidonie de la Houssaye of Franklin). Kate Chopin used the bayou as the setting for "A Gentleman of Bayou Teche," published in her short story collection *Bayou Folk* (1894). The waterway also appears in Alice Ilgenfritz Jones's "sentimental-historical romance" *Beatrice of Bayou Teche* (1895); Robert Olivier's historical-fiction trilogy *Pierre of the Teche* (1936), *Tidoon: A Story of the Cajun Teche* (1972), and *Tinonc: Son of the Cajun Teche* (1974); and Karen A. Bale and Kathleen Duey's young reader's novel *Swamp: Bayou Teche, Louisiana, 1851* (1999), among others.

More recently the Teche has figured strongly in the best-selling Dave Robicheaux detective novels. In *Crusader's Cross* (2005), for example, James Lee Burke, an author with New Iberia roots, wrote: "I could see the gardens behind the Shadows, a plantation home built in 1831, and the receding corridor of oak and cypress trees along the banks of the Teche, a tidal stream that had been navigated by Spaniards in bladed helmets, French missionaries, displaced Acadians, pirates, Confederate and Yankee gun crews, and plantation revelers who toasted their own prosperity on paddle wheelers that floated through the night like candlelit wedding cakes."[5]

Another characteristic that helps explain the Teche's allure is its cultural landscape, which outsiders have consistently viewed as exotic. From the colonial era onward, the bayou sliced through south Louisiana's French-speaking interior—an enclave occupied by an enticing amalgam of Cajuns, Native Americans, and black, white, and mixed-race Creoles. While nauseating more prudish Anglo-Saxonist observers, the region's striking *foreignness* appealed to nineteenth-century romantics and their appreciation for the gothic—found, for example, in the ominous ruins of plantation homes and sugarhouses, or in the melancholy groves of sprawling, twisted live oaks dripping with Spanish moss. Later this same quality satisfied a nostalgia for the simplicity of a pre-modern golden age—one that never really existed, particularly for enslaved blacks.

In short, observers saw the Teche and the lands adjoining it as simultaneously fertile, accessible, prosperous, beautiful, and mysterious. But, importantly, the bayou also bore the cachet of appearing in Longfellow's celebrated poem. It is true that *Evangeline* mentioned not only the Teche but also the Atchafalaya and the Lafourche. Yet Longfellow presented the Teche and its luxuriant banks as the Acadian exiles' sublime goal, their "Eden," as the poet himself wrote. Because of the immortal link between Longfellow and the Teche, today visitors find along the bayou at St. Martinville the "Evangeline Oak" and so-called "Tomb of Evangeline"; underneath the oak, overlooking the Teche, a bust of Longfellow; and a short distance upstream, the Longfellow-Evangeline State Historic Site.

Evangeline belongs to the Teche. As New Iberia songsmith Alfred Dieudonne wrote mawkishly in 1929, "Evangeline! Evangeline! You live forever on this stream!"

It is unclear, however, when the bayou came to be known as *the Teche*. Did Native Americans give this name to the bayou prior to contact with Europeans? Or did Europeans name the waterway, perhaps borrowing a Native American word for the purpose? Before standardized spelling, colonial scribes rendered the bayou's name variously, regarding no spelling as more correct than another: *Tache*, *Tage*, *Tash*, *Techa*, *Teche*, *Tesche*, *Teich*, *Teichte*, *Teis*, *Tex*, *Texe*, *Teych*, *Teyche*, *Thecte*, *Theich*, *Theiche*, *Theis*, *Theix*, *Theiz*, *Thex*, and *Tieche*. No doubt other variations exist in old documents. (The bayou also went by the name *Rivière des Attakapas*—the latter word itself having several spellings.)

The earliest known use of the word *Teche* dates to July 16, 1765. That month pioneer cattleman Jean-Baptiste Grevemberg wrote to colonial administrators in New Orleans, complaining about Acadian trespassers and requesting formal title to his *vacherie* (ranch). In reply, the Spanish colony's French administrators, Charles-Philippe Aubrey and Denis-Nicolas Foucault, granted Grevemberg title to land bounded by "*la Rivière de Teiche*."[6]

But why did anyone name the bayou *Teche* (no matter the word's spelling) in the first place? Three rival etymologies explain, or attempt to explain, the term's origin.

The most dubious of these etymologies appeared as early as 1859, when the *Opelousas Courier* newspaper authoritatively explained: "Many suppose the name of our beautiful bayou to be of Indian origin. But such, however, is not the case. The stream was called after Edward Teche, the noted pirate, who is said to have had a rendezvous on Berwick's Bay [near the mouth of the bayou]." Edward Teche, or Teach as his name is usually spelled, was none other than Blackbeard the Pirate. There is no evidence, however, that Blackbeard—who in the early 1700s terrorized the Caribbean and southern colonies of British America—ever marauded the Louisiana coast, much less on or near the Teche.

Another questionable etymology asserts that *Teche* is a corruption of *Deutsch*, the German word for "German." Dating back at least as early as 1883, this explanation holds that the French or Spanish named the bayou Deutsch because of German settlers living along the waterway, and that over time Deutsch morphed into Teche. There were indeed Germans in the area; in 1779 Spanish military officer Francisco Bouligny wrote to Louisiana governor Bernardo de Gálvez from the Teche: "I met four German families when I first crossed the Big River [the Atchafalaya] and they were in the greatest misery and unhappiness, so I told them to come here [to the Teche] and settle with the others [in my expedition]."

The Deutsch theory is problematic, however, because the number of Germans living on the Teche was always negligible. Moreover, when the French and Spanish did encounter sizable numbers of Germans in south Louisiana, they did not use the German word *Deutsch* to describe them, but, naturally, the French and Spanish words *Allemand* and *Alemán*. Indeed, a Louisiana bayou on which many German pioneers did settle still bears the name *Bayou des Allemands*. (The original name of Patterson, located at the mouth of the Teche, was variously Dutch Settlement, Dutch Town, and Dutch Prairie. This could be regarded as evidence for the Deutsche theory, since Dutch and Deutsche sound similar and are in fact related terms. On closer inspection, however, such a theory would fall apart, because Patterson's Dutch founders did not arrive on the bayou until the early 1800s, after the waterway's name had been firmly established as Teche.)[7]

A more convincing explanation is that *Teche* derives from a local Native American word of the Chitimacha language meaning "snake." This account appeared in print as early as 1922 in *The Bayou and Its People*, a promotional booklet issued by the Southern Pacific Railroad Company. The booklet referred to "the old Indian legend that the Teche came by its name because of an enormous serpent. . . ." A 1927 issue of the *Journal of American Folklore* included a more detailed version of the legend: "Bayou Teche . . . gets its name from an Indian legend, according to which the Indians came one day upon a monstrous snake, twisting and writhing and spitting forth fire; the tribe, with mighty shouts and great war cries, overcame it; this snake was the bayou, and the Indians gave to it the name 'Tenche,' signifying snake, which has now become Teche. . . ."[8]

It is certainly true that the traditional Chitimacha name for Bayou Teche translates into English as "Snake Bayou." The problem, however, is that "Snake Bayou" in Chitimachan is rendered *qukx caqaad*—neither element of which bears any resemblance to the word *Teche*. (*Qukx* means "snake" and *caqaad* means "bayou.")[9]

Where, then, did the word *Teche* come from and why the need for it? There is, for example, no word for "snake" in any other Native American language spoken in the region (Attakapas, Choctaw, Houma, Koasati) that lends itself convincingly as the source of *Teche*.[10]

Faced with this puzzle, I developed two theories that attempt to explain the origin and meaning of *Teche*. I cannot prove either is correct, so they should be viewed with skepticism. First, I propose *Teche* might descend not from the Chitimacha word for snake, but from the Chitimacha word for another writhing, elongated, legless creature, the worm. In Chitimacha the word for worm is *tcī'c* or *tcīīc*, pronounced "cheesh." This pronunciation bears some resemblance to *Teche*. Indeed, noted Chitimacha benefactor and basket collector Mary McIlhenny Bradford, when recording the name of a certain Chitimacha basket design called *tcī'cmîc*, or "worm-track," spelled it *tesh mich*, so much did *tcī'c* sound like "tesh" to her ear. But, again, the tribe's use of *qukx caqaad* to signify the bayou ultimately suggests a non-Chitimacha source for *Teche*.[11]

My second theory, therefore, is that *Teche* derives from a Native American word of Caddo origin meaning "friend"—and that *Teche* is a Spanish colonial term for "Texas." I developed this explanation after reading a Spanish colonial manuscript from 1690. That centuries-old document recorded the first encounter between Spanish explorers and the indigenous peoples of what is now coastal southeast Texas. Prepared only a year after the event it described, the manuscript stated: "As soon as the Indians became aware of our presence, they made for the wood. . . . The Indian who served as our guide himself entered the wood, and called to the others, declaring that we were friends, and that they should have no fear. Some of them—and among these was their captain—came out and embraced us, saying *'thechas! thechas!'* which means 'friends! friends!'"[12]

"*Thechas*" subsequently became the name by which the Spanish referred to the entire region, as anthropologist John R. Swanton explained:

> This word appears in the forms *texas*, *texias*, *tejas*, *tejias*, *teysas*, *techan*, etc., and hence these Indians were called Texas Indians and the word was subsequently applied to the [Mexican] province of Texas and taken over by the American colonists as that of the Republic and later State of Texas. The *x* in this word was not, however, pronounced by the Spaniards as it is in English. Sometimes it was made equivalent to [the] Spanish *j* . . . but I have usually found that in the early Spanish narratives it is employed for the English *sh*, for which the Spanish

language provides no specific sign. . . . I, therefore, believe that the original pronunciation of Texas was Tayshas, although . . . it may have been Taychas.[13]

Mexican priest and scholar José Antonio Pichardo made the same observation in the early nineteenth century, decades before Texas broke away from Mexico to become an independent nation and then an American state. As Pichardo noted, "The word Texas itself some write with an *x* and others with *ch*, Techas, and its correct pronunciation will be that which a Frenchman would give it, reading this word as if it were the French *Techas* [my italics]." Pichardo further explained that Spanish missionaries "who knew the language of those Indians [in Texas] wrote with an *x* that which the Indians pronounced with the *ch*. This is seen in their [the missionaries'] original manuscripts."[14]

Noting the similarity of *Techas* (and its alternate spellings) to *Teche* (and its alternate spellings), I propose that Spanish explorers named Bayou Teche for the Mexican province of Texas. Thus, the name *Teche* would derive not from a Chitimacha word meaning "snake," but from a word in another Native American language, Caddo, meaning "friend."

If this seems fanciful, consider that the Spanish viewed all Louisiana west of the Mississippi as Spanish territory—that is, as an extension of the Mexican province of Texas. In fact, a 1757 Spanish map wishfully showed the province of Texas stretching all the way to the Mississippi. Consider also that the Spanish capital of Texas stood from 1729 to 1770 in present-day Louisiana (near Natchitoches). Consider also that at one time or another the Spanish referred to the present-day Sabine and Mermentau rivers, located in southwest Louisiana, as the *Rio Mexicano* (Mexican River). In this context, it does not seem far-fetched to imagine the Spanish naming a major Louisiana bayou the *Techas* (Texas), which in time morphed into *Teche*.[15]

There are, however, other possible explanations. For instance, *Teche* might derive not from a Native American word, but from a French, Spanish, or even Afro-Caribbean term. (The word *gumbo*, for example—so strongly associated with south Louisiana cuisine—can be traced through the Caribbean to west Africa.) Indeed, a river in France bears the name Tèche, another the name Tech. Perhaps a French pioneer carried one of these names to the New World, christening the bayou after a homeland river? Like my *Techas* proposal, this is only a theory. Barring discovery of a "smoking-gun" document, we may never know the actual origin of the word *Teche*.[16]

As for the word *bayou*, scholars assert with authority that it comes from the Choctaw word *bayok* or *bayuk* (later rendered *bok*), meaning "river" or "stream." Most bayous are slow, muddy, and relatively small waterways that become torrents only during significant rainfalls. Although most often associated with Louisiana, the word is used along the Gulf Coast from Texas to Florida and as far north as Missouri. Regardless, it is Louisiana that bears the nickname "the bayou state." Here the word is applied frequently to rivers, streams, and sometimes even *coulées* (small water-filled gullies or ravines).[17]

Regarding other terms used in this book: Unless otherwise stated, *Louisiana* refers to the current state of Louisiana, not the entire Louisiana colony and territory that once made up about one-third of the continental United States. The term *south Louisiana* refers to the present-day Acadiana region, so-called because of its large Cajun population. When not referring to the Native American tribe of the same name, *Attakapas*, as well as *Poste des Attakapas* and *Attakapas District*, refer to a region of south-central Louisiana occupied roughly by the modern-day parishes of Vermilion, Lafayette, Iberia, St. Martin, and St. Mary. Similarly, when not referring to the Native American tribe or town of the same name, *Opelousas*, as well as *Poste des Opelousas* and *Opelousas District*, refer to a region of south-central Louisiana made up roughly by the modern-day parishes of St. Landry, Calcasieu, Cameron, Beauregard, Allen, Jefferson Davis, Evangeline, and Acadia (though in the context of this book I usually mean the section of the Opelousas poste or district closest to the Teche).

Appearing often in this book, the words *Acadian*, *Cajun*, and *Creole* also deserve explanation. In this book *Acadian* refers to eighteenth-century exiles from the Maritime Provinces of Canada (Nova Scotia, New Brunswick, and Prince Edward Island) who settled in south Louisiana, as well as to their offspring. Although deriving from *Acadian*, the word *Cajun* has a slightly different meaning, referring to descendants of the Acadian exiles and the several other ethnic groups with whom they intermarried on the semitropical frontier, such as the Spanish, French, and Germans.

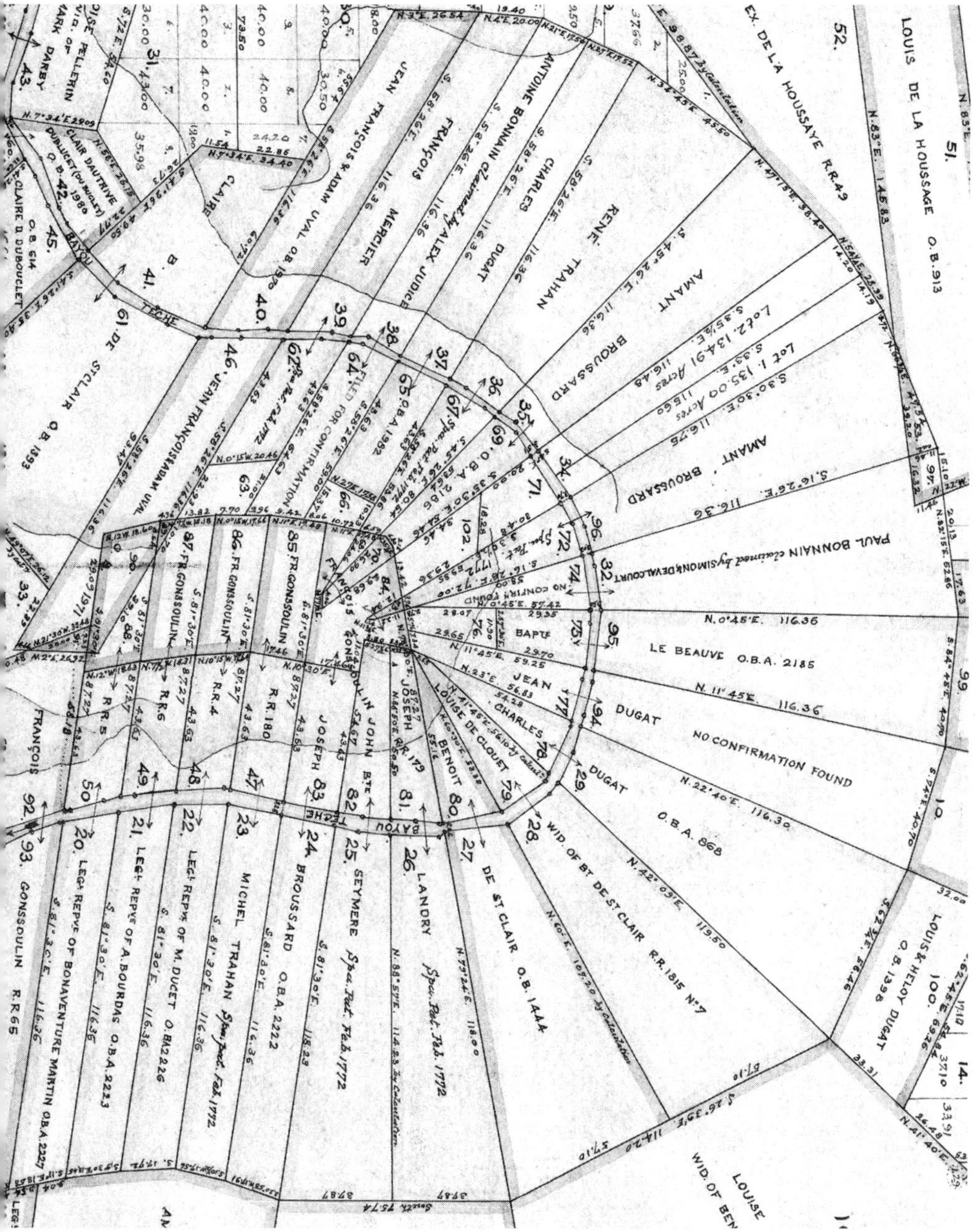

1875 land map showing parcels along the Fausse Pointe oxbow extending forty linear arpents from the Teche. Some landowners on the map received their grants in the early 1770s. Source: Office of State Lands, State of Louisiana, Baton Rouge, LA.

Indeed, I consider Cajuns an amalgam of various ethnic groups, which explains why modern Cajun surnames hail from diverse sources. For example, Guidry, LeBlanc, and Theriot are Acadian; Fontenot, Soileau, and Fuselier are French; Huval, Stelly, and Toups are German; and Romero, Viator, and Segura are Spanish. Anglo and Scots-Irish pioneers were also absorbed into the embryonic ethnic group. As legendary Cajun musician Dennis McGee once noted, "McGee, that's a French name. I don't know anyone named McGee who doesn't speak French."[18]

As for the word *Creole*, in the broadest sense it historically meant "native to the colony" or "native to Louisiana." Confusingly, however, it also referred to native Louisianians of black, white, or mixed-race heritage. A person of French or Spanish ancestry born in the colony, therefore, is called a French Creole or Spanish Creole; a mixed-race Louisianian is referred to as a Creole of Color (or, in the antebellum period, a *gen de couleur libre*, "free person of color"), who occupied a middle ground between whites and enslaved blacks; and Louisianians solely or largely of African descent are known as *black Creoles*.[19]

In reality, however, people do not fit neatly into definite ethnic or racial categories. As such, the above labels should be considered working definitions merely for the purposes of this book.

The geographic scope of this study covers a relatively small area, namely the Teche itself and a swath of land extending forty arpents back from the bayou on either side (about 1.5 miles, or 3 miles total for both banks)—*arpents* being the archaic French and Spanish unit of measurement used for both area and length. I chose forty arpents because colonial land grants often stopped this distance from the Teche; and because it was at this distance from the bayou that in 1833 the state required a public road to be maintained. Thus, in present-day New Iberia the major thoroughfare known as East Admiral Doyle Drive is the old Forty Arpent Road and sits forty linear arpents from the Teche. Likewise, another Forty Arpent Road segment (now known as Gondron Road and Harold Landry Road) still runs along the bayou on either side of Loreauville—a span historian Glenn R. Conrad called "probably the best example of a forty arpent road" in the region. In short, the forty-arpent boundary provides a convenient limit, because virtually everything discussed herein occurred either on the Teche or within forty linear arpents of the Teche.

As Darby noted in 1817, however, this long, narrow strip comprised of the bayou and its banks exerted an economic impact far out of proportion to its size. The Teche, after all, flowed "through one of the most fertile and thickly populated districts of Louisiana," as the U.S. Army Corps of Engineers observed later the same century, calling the region "the richest sugar country in Louisiana." Importantly, this fertile strip also served as an incubator for younger generations of settlers, who, like birds leaving a nest, spread out to settle other parts of Acadiana.[20]

Finally, some observers refer to the Teche and its adjacent lands as the Teche Valley—a term I avoid because, though not necessarily incorrect, *valley* to me evokes a low area between hills, ridges, or mountains, as one might find, for example, in the Shenandoah Valley of Virginia. But the Teche flows through a region so comparatively flat that *valley* seems to me an ill-chosen term. Others refer to the region as the Teche Corridor, another term I spurn because to my ear it sounds like the name of an elevated freeway. I prefer to use the term Teche Country, which in my opinion describes the region sufficiently (even if it sounds a bit touristy).

Before delving into the Teche's remarkable history, I provide here a cursory tour of the bayou. Part II of the book, however, contains a more detailed account of my own trip down the Teche by canoe.

We begin at the sleepy Cajun town of Port Barre (2010 population 2,055), which traces its origin to a colonial-era trading post. Here the Teche springs unassumingly from its mother stream, Bayou Courtableau, flowing southwest for about four miles before jutting westward in a bulbous oxbow (a U-shaped bend in a waterway). After seven miles this bend curves back on itself to arrive at Leonville (pop. about 1,085). The oxbow then continues to the northeast before arcing southward toward Arnaudville, some seven miles' distance. A rustic community (pop. about 1,060), Arnaudville straddles the border between St. Landry and St. Martin parishes. In recent years the town has become a "hip" rendezvous for artists, musicians, and cultural activists.[21]

It is important to note—and those who know the modern Teche might find this surprising—that until the early twentieth century vessels could ascend the Teche no farther than Arnaudville *except during floods*: either the bayou above the town was too low or, particularly in summer and fall, waterless, or nearly so. The construction of Keystone Lock and Dam near St. Martinville raised the upper Teche's water level, so that after mid-1913 vessels could ascend the bayou as far as Leonville. It was not until 1920, however, that dredging of the uppermost reaches of the Teche created the bayou that exists today—one navigable year-round to Port Barre and Bayou Courtableau.[22]

Three minor tributaries enter the Teche above Leonville. They are, from north to south, Bayou Toulouse, which flows into the Teche just below Port Barre; Bayou Little Teche, formerly known as Bayou Marie Croquant; and Bayou del Puent, formerly spelled Bayou del Puente—meaning "Bayou of the Bridge" or simply "Bridge Bayou."

Bayou del Puente appears in historical documents as early as 1812. The name is Spanish and so perhaps a remnant of Spain's nearly four-decade rule of Louisiana. As for Bayou Little Teche's previous name, Bayou Marie Croquant, it has been rendered *Bayou Marie Crocan* and *Bayou Maricoquant*, and during the Civil War it was called *the Barri-Croquant*. As one historian noted, a Union general "committed a minor error which was to baffle civil war scholars for generations": "On all the Union maps, the letter 'M' in Marie Croquant was so blurred as to render interpretation all but impossible. [General] Franklin read it as a 'B,' possibly thinking of Barre [as in 'Port Barre'], and all Union correspondence thereafter refers to the bayou as the Barri-Croquant."[23]

Yet even that tributary's current name, Bayou Little Teche, can be confusing (at least for the historian), because the uppermost stretch of Bayou Teche was itself sometimes called the Little Teche. In 1886, for example, a U.S. Army Corps of Engineers officer commented, "From here [Leonville] up the stream is called the Little Teche. . . ."

More puzzlingly, early-nineteenth-century land maps sometimes referred to the Marie Croquant as the main channel of the Teche and showed it winding toward and under the town of Washington—a town otherwise never associated with the Teche—to become what is today Bayou Carron. Worse, some historical documents referred to these same waterways as "Bayau [*sic*] Catereau or Teche" or "Bayou Grand Louis or Teche, or Carron"—indicating uncertainty or disagreement over what to call these tributaries, none of which, in any event, should be confused with the Teche proper that is the subject of this book.[24]

About two miles south of Arnaudville the Teche bulges notably in another oxbow, or rather two consecutive oxbows, each protruding in the opposite direction of the other. First the bayou juts impulsively to the east, then on turning back overcorrects, and finally resolves on a sensible middle passage to the south. Along this erratic stretch the Teche runs through Cecilia (pop. 1,980), flows underneath Poché Bridge (a landmark known chiefly for the *boudin* sold at a nearby market), and arrives at Breaux Bridge (pop. about 8,140). Known in earlier days as *Pont Breaux*—a name still used by some as an affectionate term for the town—Breaux Bridge has recently gained, like Arnaudville, a "Cajun chic" aura. Composed of vintage brick storefronts, its downtown boasts restaurants, antique shops, and a small but vibrant bohemian population.

From Breaux Bridge the bayou shoots south-southeast for about four miles before turning eastward for over two miles to reach Parks. The most prominent feature of this small community (only about 655 residents) is, fittingly, its large public park. (My fellow researcher, Donald Arceneaux, suggests Parks is so named because during the colonial era it held a number of cattle pens—*les parcs* in Louisiana French—used to corral beeves before local French-speaking cowboys drove them across the Atchafalaya to New Orleans slaughterhouses.)[25]

Below Parks the bayou heads generally south, after about four miles reaching the Louisiana Sugar Cane Co-op mill, which many still call by its old name, Levert-St. John. From October to January this refinery and others farther down the Teche—at New Iberia, Sorrel, and Franklin—hiss like lurking dragons, emitting great columns of roiling steam as powerful rollers crush sugarcane taken from surrounding fields. The extracted saccharine juice is boiled and, ultimately, converted into molasses and white granulated sugar. Planters today share a handful of refineries, while in former times each plantation operated its own "sugar house." These old sugar houses did not manufacture white sugar, but a rich, sticky brown sugar seeped in molasses. Some early planters, however, converted their sugar crops into rum, called *tafia* in Louisiana French.

Levert-St. John might be said to mark, at least symbolically, the inexact border between the historically sugar-growing and cotton-growing lands along the Teche. As a journalist heading up the bayou noted in 1870, "As I went up the Teche . . . it became less desirable for sugar and more desirable for cotton; and when I got to New Iberia, and for twenty miles further north, the plantations were about equally divided between sugar and cotton—some growing both on the same plantation. As the one runs out the other runs in; but where both are grown it is really not as good for either."

While perhaps unclear to a boater, from this point southward the Teche often passes through one present-day sugarcane plantation after another. The slaves, freedmen, sharecroppers, and tenant farmers of earlier days no longer exist, but the plantations endure—consolidated, mechanized, corporatized. However, one can still occasionally glimpse an antebellum "big house" on the edge of the bayou. Each once served as the center of social and business life on a sugar plantation. These houses include, among others, Shadows-on-the-Teche, Bayside, Albania, Oaklawn Manor, Shadowlawn, Arlington, Frances, and Bocage-on-the-Teche—as well as St. John, the "big house" at Levert-St. John. (From the bayou the house sits to the refinery's right, its simple elegance contrasting starkly with the byzantine sprawl of the next-door industrial site.)[26]

About a mile south of the Levant-St. John refinery lies St. Martinville. Although outsized by present-day Breaux Bridge, St. Martinville (pop. about 6,115) was once the northernmost of three major shipping hubs on the Teche—the others being New Iberia and Franklin. Here, in the shadow of St. Martin de Tours Catholic Church, resides the epicenter of the Evangeline myth.

As historian Carl A. Brasseaux has pointed out, Evangeline never existed, nor did Emmeline Labiche, identified by dubious sources as Longfellow's "real life" model for Evangeline. Neither woman, then, waited for a lost love under the Evangeline Oak, as legend claims. The nearby Tomb of Evangeline is in reality a cenotaph to the memory of two fictional characters. A short walk from the tomb, however, stands the Acadian Memorial, housing a Wall of Names that lists some three thousand actual Acadian exiles who found refuge in colonial Louisiana. In the memorial's courtyard burns the eternal flame, around which is inscribed the motto, "*Un peuple sans passé est un peuple sans futur*"—a people without a past is a people without a future. A few doors down is the Memorial Museum and, next to it, a museum dedicated to African Americans of the Attakapas region in the eighteenth and nineteenth centuries.[27]

From St. Martinville the Teche wobbles to the south and after less than four miles reaches Keystone Lock and Dam, built, as mentioned, to raise the water level of the upper Teche. About two miles below Keystone the bayou flows into a large oxbow known as Fausse Pointe (False Point). Stretching eastward, this bend runs about twelve miles in length, flowing along the village of Loreauville (pop. about 890) and passing the banks where the Cajuns' brutalized ancestors, the Acadian exiles of Nova Scotia, settled beginning in 1765. The bayou then twists westward to arrive at New Iberia, founded by the Spanish in 1779 as Nueva Iberia (so named for the European peninsula occupied partly by Spain). New Iberia is today the largest town on the Teche (pop. about 30,620) and has been called "the Queen City of the Teche" since at least 1899, when a St. Martinville newspaper tweaked the rival downriver town by observing, "New Iberia is not as progressive as it thinks it is, even if they call it the queen city of the Teche." Only a block from the bayou, in downtown New Iberia, sits the Bayou Teche Museum, whose exhibits feature the history, peoples, and industries of the waterway.[28]

From New Iberia the bayou runs southeastward, reaching Jeanerette (pop. 5,530) after about eleven miles. Here on the banks sits the Jeanerette Museum, an intriguing hodgepodge of local ephemera, including hand-carved wooden mold patterns once used to make iron gears at the adjacent Moresi Foundry. Established in 1852, the foundry is still in operation today, albeit with the help of high-tech gadgets like robot welders. (The now defunct Lutzenberger Foundry stands upstream at New Iberia; its premises have been examined by archaeologists and historians, including myself. We had hoped to find there the site of colonial Nueva Iberia—but no such luck.)[29]

About three miles beyond Jeanerette the Teche passes through a crossroads called Sorrel (pop. about 765). There on the bayou stands the massive St. Mary Sugar Co-op refinery. In front of the co-op, along Highway 182, stands a neglected state historical marker, informing passersby about eighteenth-century cattle baron Jacques Joseph Sorrel, whose sizable Teche-side *vacherie* (the French version of the sign actually reads "*ferme d'élevage*," or breeding farm) once stood on or near the site of the present-day refinery.

Four miles past Sorrel the Teche encounters a double oxbow, the first known historically as Indian Bend, the second as Irish Bend. From start to finish the two oxbows run about nineteen miles. At the first oxbow's tip, which juts sharply northeast like the horn of a rhino, sits the Sovereign Nation of the Chitimacha, the ancestral lands of the federally recognized Chitimacha tribe of Native Americans. Here in the town of Charenton (pop. about 1,900) the tribe operates a museum as well as a casino with a hotel, restaurants, and occasional "big-name" entertainment. Most of the tribe's ancestral land lay on the bayou's west bank; some of it, however, sits on the east, meaning boaters, if only briefly, pass completely inside tribal lands.

At the end of this oxbow sits Baldwin (pop. about 2,435), near whose site the Teche Country's earliest known European settler, André Masse, dwelt toward the end of his life in the late eighteenth century. On the far side of Baldwin boaters follow either a thousand-foot stretch of the Teche's original course, now almost entirely clogged with vegetation, or an 1,800-foot detour, which heads south into the lower Charenton Drainage and Navigation Canal, toward West Côte Blanche Bay, before swinging northeast around a small wooded isle to reunite them with the Teche.

Sweeping northeast, the bayou enters the second oxbow, Irish Bend, so named because three families of Irish ancestry, the Murphys, the Cafferys, and the Porters, settled here around 1820. Sugar planter Alexander Porter purchased the tip of the oxbow and there built Oaklawn Manor. The antebellum home still exists, can be seen from the bayou, and is presently the home of former Louisiana governor Mike Foster. It was on this oxbow that the Battle of Irish Bend occurred in 1863 between nearly ten thousand Union and Confederate troops.[30]

From Oaklawn Manor the Teche turns south, then whips back to the west to complete the second oxbow. At its end sits the Sterling sugar refinery—now owned by the Patout family of nearby Patoutville—and the town of Franklin (pop. 7,660), renowned for its elegant old homes and fluted, triple-globed street lamps, each bearing on its base the embossed words "DO NOT HITCH" (a vestige of the town's horse-and-buggy days). Looming over the Teche, the contemporary St. Mary Parish courthouse seems to mock Franklin's Old South charm. As a history of Louisiana courthouses tactfully observed, "The structure's modernistic design is unfortunately not in keeping with the parish's rich architectural heritage." An acquaintance of mine, however, once described the structure with typical Cajun frankness, declaring it looked "like the backside of a refrigerator."[31]

About two miles beyond Franklin the bayou reaches Garden City, which is not a city at all, nor even a town, but a cluster of dwellings (pop. about 55). Its former post office, which still stands, briefly appeared in the groundbreaking 1960s film *Easy Rider*, along with sections of Franklin—including the present courthouse, which at the time of filming was under construction.

Two miles farther downstream the Teche passes Centerville (pop. about 550), so named supposedly because the town lay at the center of St. Mary Parish (which is not really the case, even taking into account the parish's original shape). About six miles farther downstream sits the communities of Ricohoc and Calumet, both of which take their names from old sugarcane plantations. These two clusters of rural homesteads are today divided by a huge channel known as Wax Lake Outlet or, as some call it, Calumet Cut. This manmade waterway—spanning nearly six hundred feet from bank to bank and running about sixteen miles from its headwaters in the Atchafalaya swamp to its mouth at Atchafalaya Bay—cuts ignominiously right through the course of the Teche, whose waters are often sealed off from the outlet by massive floodgates.

On the Calumet side of Wax Lake Outlet the Teche resumes its course, shortly reaching the town of Patterson (pop. about 6,110), an inland port home to a fleet of commercial vessels serving the offshore petroleum industry. Here the Teche ends, flowing into the Lower Atchafalaya, down which sit the towns of Bayou Vista (pop. about 4,650) and Berwick (about 4,945).[32]

Earlier historical maps show the Teche continuing past Patterson to flow into Berwick Bay at present-day Berwick. Some even show the

Teche running concurrently with the Lower Atchafalaya below Berwick, both occupying the same channel to eventually pour into Atchafalaya Bay and, beyond it, the Gulf of Mexico. More recent and present-day maps, however, agree the Teche ends at Patterson. The U.S. Army Corps of Engineers and the Cartographic Information Center at Louisiana State University, among others, subscribe to this view, one I at first rejected. The more I researched the question, however, the more I came to believe that the Teche did indeed end at Patterson. Conversations with retired offshore oilfield fleet owner, nautical historian, and octogenarian Fulton C. "Butch" Felterman of Patterson—who once stated "it pained him greatly to hear the . . . Lower Atchafalaya River called Bayou Teche"—hastened my conversion. So did a timely proclamation by the City of Patterson, which in March 2014 officially declared that the waterway between Patterson and Berwick "shall forever be referred to in the official records of the City of Patterson as the Lower Atchafalaya River." I am not averse, however, to others regarding the Lower Atchafalaya as a sort of "honorary" extension of the Teche. For the purposes of this book, however, I will consider the mouth of the Teche to be at Patterson, and the waterway below it to be the Lower Atchafalaya.[33]

Part One

1

Settling the Teche

Like much of present-day Louisiana, Bayou Teche owes its existence to the Mississippi River. Not only does the modern bayou twist its way through a rich alluvial plain deposited millennia ago by the Mississippi, it follows a path carved out by the big river some 3,800 to 5,500 years ago. When the Mississippi withdrew eastward, the Red River occupied the abandoned channel. Then about 2,000 years ago the Red River changed its course, leaving behind the slow, muddy waterway now known as Bayou Teche. This geologic sequence explains why the modern Teche has three natural levees—its own present-day levee; a steep, narrow inner relic levee of ruddy soil deposited by the Red River; and a wide, gently sloping relic levee of brown to gray alluvium left by the Mississippi.[1]

Archaeological evidence indicates that Native Americans of unknown tribal affiliation (called Paleoindians) arrived on the prairies west of the Teche region some 13,500 years ago—long before the Teche flowed through south Louisiana. Some 4,500 years ago other Native Americans constructed an earthen mound at Avery Island, less than ten miles from the present-day bayou. The four parishes through which the Teche runs—St. Landry, St. Martin, Iberia, and St. Mary—today boast seventy-four known mound sites, mostly ranging in age from 500 to 1,300 or more years old. Not all these mound sites stand on the bayou, but some do, and from the water a trained eye can easily discern them.[2]

By the dawn of the colonial era three Native American tribes lived on or near the Teche: the Opelousas, the Attakapas, and the Chitimacha.

While little is known about the Opelousas—anthropologist Swanton required less than a page to convey his total knowledge of the tribe—more is known about the Attakapas, the tribe that eventually gave its name to much of the region bisected by the Teche. French and Spanish explorers feared the Attakapas, whose name in Choctaw, justly or not, meant "man eater." More than one circa 1720 map reflected this fear: across south Louisiana they displayed the warning "*Indiens errans et antropophages*"—wandering Indians and man-eaters. According to a Chitimacha source, the Attakapas once inhabited a village on the Teche at present-day Loreauville. But the tribe had forsaken the bayou by the mid-1700s, instead living on Bayou Vermilion several miles to the west and selling any remaining claims on the Teche to French-speaking colonists.[3]

The Chitimacha, however, had long dominated the Teche, or at least its lower stretches. There the tribe erected mounds or occupied preexisting mounds. For example, a mound complex exists on the east bank of Bayou Teche near Patterson. It consists of three mounds, the largest of which today rises to a height of about nine feet and extends roughly 100 feet across, and a large refuse heap (midden) of clam shells, whose contents the Native Americans consumed as part of their regular diet. Excavations at the site indicated a series of occupations beginning around 700 to 1200 A.D. and ending with a later Chitimacha village. The present-day Chitimacha Tribe of Louisiana calls this abandoned village *Qiteet Kuti'ngi na'mu*.[4]

Several other prehistoric Chitimacha villages stood on or near the Teche. *Wai't'inîmc*, for example, sat along the bayou at Irish Bend; *Cā'tcnîc*, on the site of present-day Jeanerette; and *Okû'nkîskîn*, at an unspecified site on the Teche (probably at a sharp bend, because the name means "Deep Shoulder"). A village of forgotten name stood at present-day Baldwin; and *Nē Pinu'ne* overlooked the bayou two miles upstream from present-day Charenton, which itself sits on or near a village site called *Tcāt Kasi'tuncki* or *Caqaad kaskec na'mu*. The Chitimacha resided at this last site when the first European settlers and African slaves penetrated the region; the present-day tribe remains on that site today.[5]

European pioneers encountered these villages—whether inhabited or abandoned—during their explorations of the Teche Country. A former British spy, Thomas Hutchins, thus recorded in the 1780s: "In ascending the Tage [Teche] River, it is 10 leagues from its mouth to an old Indian village, on the east side, called Mingo Luoac. . . . [Three and a half leagues] further up, on the east side, is the village de Selieu Rouge." Hutchins misunderstood, for Mingo Luoac (also called Fire Chief) and Selieu Rouge (actually Soulier Rouge, or "Red Shoe") were not place names, but the names of Chitimacha chiefs who lived in those villages. Indeed, Spanish military officer Francisco Bouligny wrote from the Teche in 1779 of "the Chitimacha whose leader is Soulier Rouge." Mingo Luoac's village has never been identified—it could have been *Qiteet Kuti'ngi na'mu*—but that of Soulier Rouge was evidently *Tcāt Kasi'tuncki* (*Caqaad kaskec na'mu*).[6]

It is unclear when the Chitimacha along the Teche first encountered European explorers. Chitimacha oral history, however, includes numerous stories about Spanish conquistadors raiding the lower bayou for prisoners. A modern Chitimacha journalist has chronicled the following account of conquistadors assailing the tribe on Grand Lake (sometimes called *Lac Chitimacha* in colonial times), only about two miles from the Teche:

> It was here that the *conquistadores* landed, as local tradition has it. And the Chitimacha told them that they couldn't come to shore. The *nata* [chief], I like to think of him as the brave warrior who probably stood on these shores and jabbed the butt of his spear into the ground. And he looked out at these people with these huge sails and the shiny helmets and the strange beards. And he told them, "Don't come ashore." . . . And the Spaniards tried to come with guns and swords, and we beat them back.[7]

Although colonial documents do not chronicle this incident, they do indicate that Spanish explorers Alvarez de Pineda and Álvar Núñez Cabeza de Vaca sailed just below the mouth of the Lower Atchafalaya in 1519 and 1528, respectively. Neither explorer, however, actually set eyes on the Lower Atchafalaya, much less on its upcountry tributary, the Teche.[8]

Hernando De Soto's ill-fated expedition of the early 1540s came close to the Teche. After a punishing overland march from Florida, De Soto reached Louisiana only to die there or in adjacent Arkansas. The tattered remnants of his expedition, now sailing in makeshift boats under Luis de Moscoso Alvarado, skirted the mouth of the Lower Atchafalaya. Although too far from shore to see the outlet, the Spanish nonetheless observed the Lower Atchafalaya's strong freshwater current. "With a favorable wind they sailed all that day in fresh water, the next night, and the day following until vespers," a Portuguese soldier on the expedition recorded in third person, "at which they were greatly amazed; for they were very distant from the shore, and so great was the strength of the current of the river [i.e., the Lower Atchafalaya], the coast so shallow and gentle, that the fresh water entered far into the sea."[9]

Over two centuries passed, however, before a European colonist overcame the collective fear of the Attakapas and settled on the Teche. He was a Frenchman named André Masse. Born around 1700, Masse was rumored to hail from a prosperous family of Grenoble. In his younger days he allegedly served as a "chancellor" of that city (a nebulous title presumably of some distinction) and a military officer. Sailing to south Louisiana, however, he chose to reject civilization and make a frontier home along the Teche. And he did this at a time when no other white men are known to have lived along or anywhere near the bayou.[10]

Arriving in the Teche County as early as 1746, Masse established a cattle ranch on the banks of the bayou. Precisely where is unknown, but in later life he moved his ranch downstream. Former British spy Hutchins, writing around 1780, noted the location of this second *vacherie*. "Mons. Mass[e]," he observed, lived on the bayou twelve leagues (about thirty miles) from its mouth. This placed Masse's ranch near the present-day town of Baldwin. A quarter century later, New Orleans cartographer Barthélémy Lafon, apparently unaware of Masse's death around 1785, likewise fixed Masse's homestead near Baldwin—more precisely, on a bend where the Teche now flows into the lower stretch of the Charenton Drainage and Navigation Canal.[11]

The Spanish in nearby Texas likewise followed Masse's whereabouts because they viewed him as a trespasser. To them the Mexican province of Texas stretched across Louisiana to the west bank of the Mississippi River. This meant the Teche fell in Spanish territory. Yet the French in New Orleans claimed the same territory. As a result, Masse the

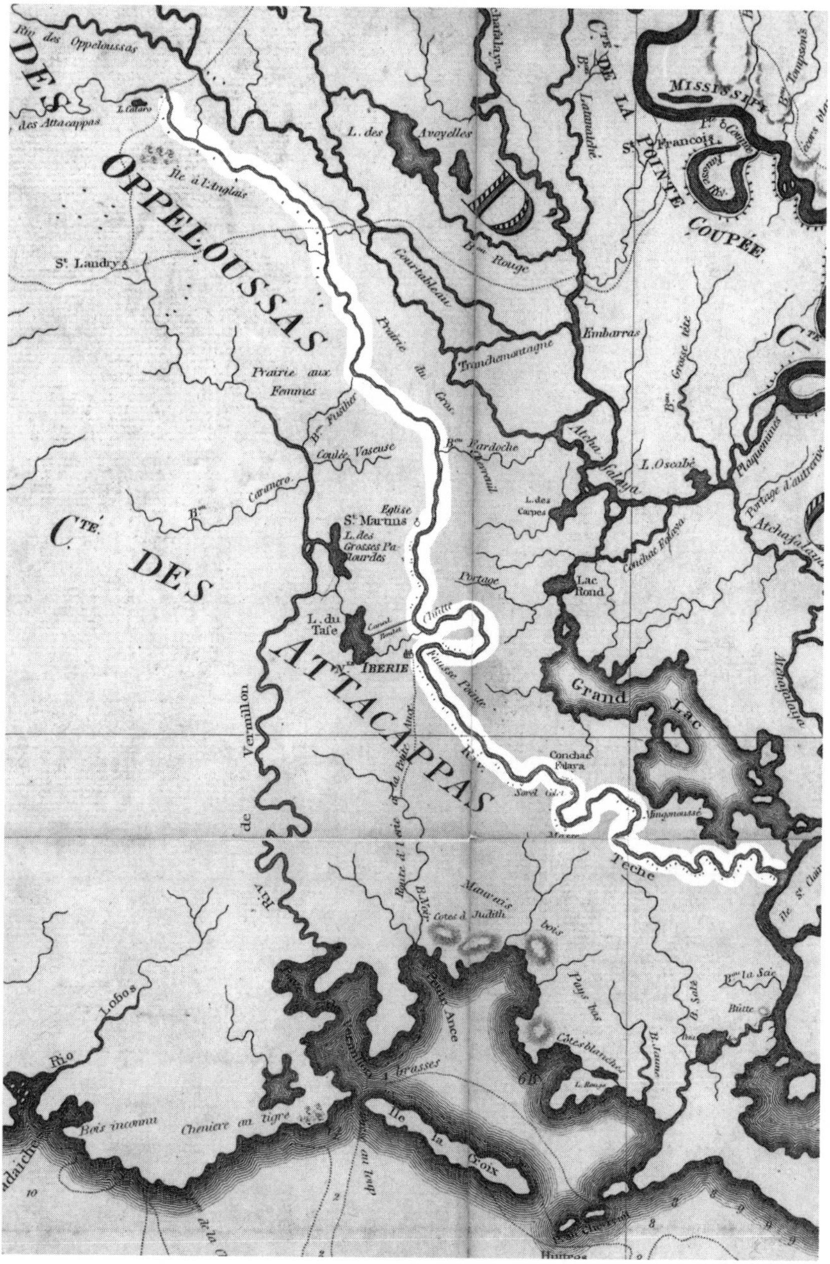

Detail of Barthélémy Lafon's 1806 map titled *Carte générale du territoire d'Orléans*, perhaps the earliest printed map to depict the Teche with any accuracy—even though Lafon showed the uppermost stretch of the bayou running off to the northwest instead of intersecting Bayou Courtableau. Source: facsimile, author's collection.

frontier cattleman (and frontier contraband runner, as some regarded him) lived unmolested in this disputed region claimed by rival Bourbon empires reluctant to anger each other. Watching from a distance, the Spanish could only gather sporadic intelligence on the audacious Frenchman.[12]

In 1757, for instance, Spanish explorer Bernardo de Miranda included Masse's dwelling (labeled "Casa de Monsiur Mas") on a map submitted to the Viceroy of Mexico and the King of Spain. (Miranda exaggeratingly depicted Masse's abode as a castle-like structure, complete with battlements.) Masse had only slaves for companions, observed Miranda in an accompanying report, and raised horses, cattle, corn, beans, and tobacco—shipping his finest tobacco to New Orleans, while the rest he sold to Native Americans or "wasted on Negroes."[13]

Even high-ranking Spanish administrators took an interest in Masse. The governor of Texas, Ángel Martos y Navarrete, complained, "Monsieur de Masse is settled on a *rancheria* with a few Negro servants . . . without any other authority or title than that of [his] own making." To avoid conflict with the French in New Orleans, the governor explained, "I have attempted to ignore this intrusion." Navarrete's predecessor, Jacinto de Barrios y Jauregui, had earlier informed the viceroy that Masse owned seven hundred head of cattle, a hundred horses, and twenty slaves (which if true made the Frenchman extremely wealthy). In addition, Barrios expressed the anxious belief that Masse exerted "absolute dominion" over a number of Native American tribes, including the feared Attakapas. Whether an exaggeration or not, the notion imparted a frightening, almost mythical quality to Masse—the white man who ruled the ferocious, even reputedly man-eating natives.[14]

Masse's friendship with the Native Americans offended the racial sensibilities of Spanish cleric José Antonio Pichardo. Writing in the early 1800s, Pichardo, like Lafon, seems to have been unaware of Masse's death years earlier. Indeed, Pichardo cursed the pioneer in the present tense, calling him a "worthless and despicable little Frenchman . . . [who] should, because of a most sordid interest, go to live in a hut among the barbarian Indians. . . ." He further condemned Masse as a "profligate Frenchman" who committed the intolerable sin of "mingling with the Indians, nourishing their vices, embracing their

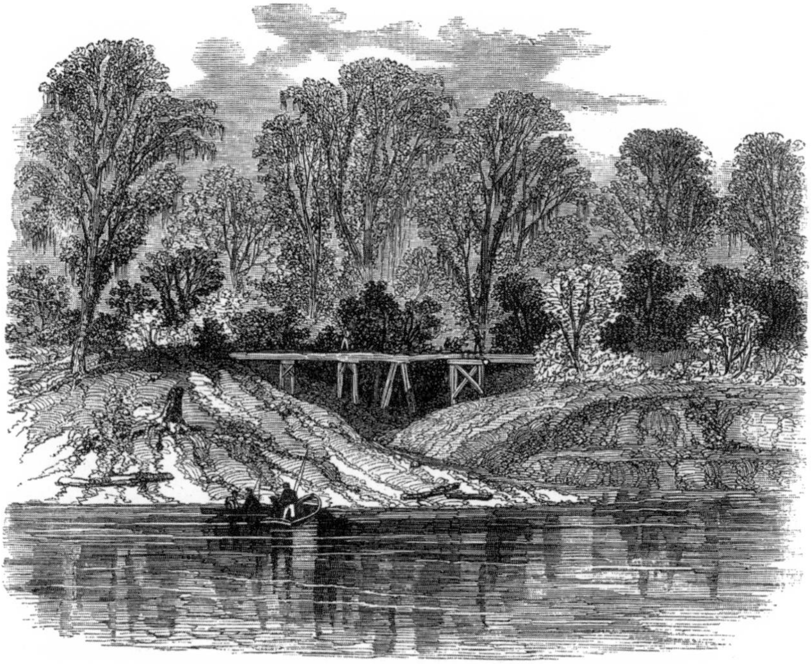

The headwaters of the Teche at Port Barre. The waterway in the foreground is Bayou Courtableau. Note the head of the Teche sits several feet above the surface of the Courtableau and is completely dry. This is how the uppermost Teche appeared for much of each year until dredged in 1920. Source: *Frank Leslie's Popular Monthly* III (June 1877).

perverted habits and customs, [and] adapting himself to their barbarous manner of living, dressing, and eating. . . ."[15]

French traveler C. C. Robin provided a more sympathetic description of the deceased Masse. As Robin recorded in his 1807 book *Voyages dans l'intérieur de la Louisiane*: "His dwelling was a simple cabin, open to the air. He slept on a bearskin, stretched on boards. He was dressed from head to foot in buckskin. His only eating utensils were a knife and a horn spoon hung at his belt. He lived thus for twenty years in the wilderness, offering hospitality to all who asked for it for as long as they wished. However, the number of parasites living on him was never large; his austerities drove them away."[16]

As for his slaves, observed Robin, Masse treated them with such leniency that "it might be said he was rather the father than the master of them." He eventually emancipated six families, granting them not only their liberty, but gifts of costly livestock. As Robin reported, Masse's freedmen "form today, at the lower end of the Teche, a little community, as indolent as in the time of their master."[17]

As years passed other settlers joined Masse, his slaves, and the local Native Americans along the murky, winding bayou. Among those settlers was Joseph Le Kintrek dit Dupont, who with his business partner, French adventurer Joseph Blanpain, traded horses, fur, and other goods with the Attakapas and Opelousas tribes. While Blanpain ended his days in a Mexican prison—the Spanish captured him trading illegally with the Orcoquiza tribe of east Texas—Le Kintrek settled at the headwaters of the Teche around 1750 and opened a trading post.[18]

There, at the future site of Port Barre, the Teche's often dry channel sprang from its less storied mother stream, Bayou Courtableau. That waterway bore the name of Le Kintrek's son-in-law, Jacques-Guillaume Courtableau. In 1765 Courtableau took ownership of eight thousand acres bordering the Teche, the Courtableau, and the minor waterway Bayou Marie Croquant (now called Bayou Little Teche), which emptied into the Teche about five miles east of present-day Opelousas. This concession made Courtableau one of the largest early landowners along the Teche.[19]

Another early landowner on the bayou, Gabriel Fuselier de la Claire, arrived in New Orleans in 1748 to conduct business on behalf of Lyons textile merchants, including his father. Fuselier chose to stay in the colony and in 1760 purchased a vast tract from the Attakapas chief Kinemo (sometimes spelled Rinemo), or "Lemonier" as the French called him. This property ran from the west bank of the Teche to the Bayou Vermilion and occupied an area of approximately seventy-five square miles.[20]

More settlers came to the Teche after France chose to cede the unprofitable colony to Spain in 1762. As it disengaged from Louisiana, the French military offered retirement and generous land grants to colonial officers in lieu of costly transport back to Europe. One of these officers, Louis Pellerin, accepted a sizable concession (almost eight square miles) along the Teche north of present-day Leonville. Another officer, the Acadian-born Joseph Deville Degoutin—a former *mousquetaire* (musketeer) who came to Louisiana of his own volition years before his exiled compatriots—received a concession of over two thousand acres across the bayou from Fuselier de la Claire. Antoine Bernard Dauterive, another officer, received a similarly large concession on both sides of the Teche in present-day St. Martin Parish. (Spanish administrators would shortly reclaim these massive eleventh-hour French concessions for their crown, leaving the retired French officers with smaller yet still significant tracts.)[21]

An absentee landowner, Dauterive resided on a plantation along the Mississippi River in the Iberville District, some thirty-five miles across the Atchafalaya swamp from his bayou-side concession. Born to a French army major, Dauterive joined the military as a youth and rose to the rank of captain. He achieved this respectably high rank despite his reputation for being "anti-social, insubordinate, and hot-headed"—to quote one of his superiors. In 1765 Dauterive and his business partner—none other than the intrepid André Masse—used their property along the Teche in a novel entrepreneurial scheme. It hinged on two seemingly incongruous elements: A herd of semi-domesticated Spanish longhorn cattle, and approximately two hundred French-speaking Acadian exiles who had come to south Louisiana in search of a new homeland. These exiles and their offspring would soon begin to intermarry with other ethnic groups on the frontier and evolve into a vital new ethnic group: the Cajuns.[22]

Many early Acadians hailed from the Centre-Ouest region of France, located on the Atlantic Coast around Poitou. They left for the New World in the early 1600s to escape the combined miseries of religious warfare, drought, famine, disease, and excessive taxation under feudal overlords. Settling in the present-day Maritime Provinces of Canada, they called their new home Acadia (*Acadie* in French) and lived for over a century on the frontier as farmers, fur trappers, hunters, and fishermen. The Acadians, however, occupied land strategically crucial to the incessantly warring British and French empires. In 1710 the British permanently seized Acadia, renamed it *Nova Scotia* (New Scotland), and forty-five years later began to expel the Acadian population by force. According to some estimates, of the fifteen to eighteen thousand Acadians who lived in pre-expulsion Nova Scotia, as many as ten thousand died from disease, starvation, neglect, exposure, and violence at the hands of the British.[23]

Of the survivors, about 3,000 exiles made their way between 1764 and 1788 to Spanish-held Louisiana, whose administrators wanted non-English settlers as buffers against British encroachment. Among the exiles to reach Louisiana was a group of about two hundred led by Acadian guerrilla leader Joseph Broussard dit Beausoleil. Administrators gave Beausoleil and his fellow exiles muskets, tools, seed, building materials, and a substantial supply of food, and sent them to live in the region called Poste des Attakapas. They intended to settle the Acadians on the east bank of Bayou Teche near the site of present-day St. Martinville. There the Acadians had agreed, in writing, to engage in Dauterive and Masse's entrepreneurial scheme: to raise longhorn cattle as a source of food for the colony. Many considered this agreement, known today as the Dauterive Compact, as vital to lower Louisiana, which had recently lost its primary source of beef, Mobile, to the British as a prize of war.[24]

To reach their new homeland on the Teche, the exiles followed an officially appointed guide, French military engineer Louis Andry, up the Mississippi to a point about twelve miles below Baton Rouge (newly garrisoned by the British). There they turned southwest onto Bayou Plaquemine, which led them into the heart of the Atchafalaya swamp. This massive forested wetland was a sodden labyrinth, each prospective path through it infested with alligators, poisonous snakes, and disease-carrying mosquitoes. On its far side to the west the Acadians found the meandering Teche.[25]

Shortly after they arrived on the bayou the Acadians apparently reneged on the Dauterive Compact. Seeking to spread out and settle on individual tracts, they instead headed a few miles south toward the large oxbow in the Teche known even then as "*la fausse pointe*" or, as it was also called at the time, "*presqu'isle*"—the peninsula. There the exiles settled in three groups "one next to the other," as a colonial source recounted, and called these settlements *le camp appelle Beau Soleil* ("camp Beausoleil"), *le premier camp d'en bas* ("the first camp down below"), and *le dernier camp d'en bas* ("the last camp down below"). The Spanish administrators of the colony had their own names for these camps. *Le camp appelle Beau Soleil* they called *quartel de le cano de tortugas* ("quarter of the channel of tortoises"); *le premier camp d'en bas* they called *quartel de la manque* (*la manque* meaning literally "the want" or "the need," but perhaps better translated here as "the break" or "the gap"); and *le dernier camp d'en bas* they called *quartel de la punta* ("quarter of the point"). These camps are thought to have stood on either side of present-day Loreauville, between the modern communities of Daspit and Belle Place.[26]

The exiles not only refused to live on Dauterive and Masse's property, they also decided against raising cattle for the two Frenchmen—the very essence of the Compact—choosing instead to buy cattle from a local pioneer landowner of Flemish descent, Jean-Baptiste Grevemberg dit Flamand. The Acadians shortly angered Grevemberg, however, by squatting on his property. As it turned out, Grevemberg himself had no clear title to the land, prompting him to write to New Orleans to plead for confirmation of his claim—the reply containing the earliest known use of the word *Teche* (albeit spelled *Teiche*). New Orleans mollified Grevemberg by confirming his Fausse Pointe concession; yet the Acadians remained on the disputed land, and in little more than five years gained formal ownership.[27]

An unidentified epidemic, possibly smallpox, typhoid, or yellow fever, struck the exiles shortly, killing about thirty-five to forty. The Acadians recovered, grew in number, and spread out along the Teche. Many claimed royal lands up and down the bayou, acquiring titles to these tracts only after Spain took formal control of the colony from French caretakers in 1769. Between 1771 and 1772, for example, Spanish governor Luis de Unzaga awarded the exiles over two dozen land grants along the Teche. These grants averaged about 415 acres each and totaled more than 10,000 acres. In coming years other Acadian exiles arrived and settled on the Teche, prompting Unzaga and his successor, Bernardo de Gálvez, to award at least another ten grants by 1783, raising the total size of Acadian land holdings on the bayou to over 18,000 acres.[28]

On these tracts the Acadians helped establish cattle ranching along the Teche, driving their herds to slaughter in New Orleans along several primitive trails. These pushed over or through the Atchafalaya swamp to reach the Mississippi and the city. For example, the Collet trail—presumably named after the Prevost dit Colette family, prominent Teche landowners—followed an overland route that used the natural levees of bayous Teche, Black, and Lafourche to reach New Orleans. Other trails, however, required herds to embark on vessels called "cattle barges" or "round boats" (supposedly "because they made the trip 'around' by the way of the Atchafalaya, Red, and Mississippi rivers to New Orleans").[29]

After cattle ranching came the first significant cash crop on the bayou—indigo, a plant used to make a blue dye valued for coloring textiles. Manufacturing the dye involved a complex, labor-intensive procedure. Enslaved workers harvested indigo plants and placed them in a series of vats to ferment. Dark blue grains formed in the increasingly putrid water, sinking to the bottom of the vats. Slaves packed this sediment into bags to dry and then cut the finished dye into squares for shipment to overseas markets. It was punishing work, in large part because it exposed the slaves to a highly toxic, potentially deadly chemical byproduct.[30]

Large-scale indigo cultivation came to Louisiana in the 1720s, when colonists set up indigo plantations along the lower Mississippi River. The plantations thrived, and three decades later indigo plantations appeared in the newly settled Teche Country. One indigo planter along the Teche was a Frenchman named Etienne-Martin de Vaugine de Nuisement. Despite his noble-sounding surname, de Vaugine hailed from a middle-class family in Vosves, France, a hamlet now inside the modern city of Dammarie-lès-Lys, located on the Seine about thirty miles southwest of Paris. Choosing a military career, de Vaugine enrolled in the royal artillery by 1742, when the War of the Austrian Succession erupted across Europe. He saw action at the Siege of Breisgau (in present-day southwest Germany), and at the Siege of Tournai and the Battle of Fontenoy (both in present-day Belgium).[31]

Serving later in Genoa and Corsica, de Vaugine arrived in Louisiana in 1751 as a lieutenant in the French infantry. By the early 1770s, however, he had transferred his allegiance to Spain, which by then held the former French colony. Commissioned a captain, de Vaugine received a roughly 2,000-acre land grant in 1772 from Spanish governor Luis de Unzaga. This property sat on the Teche midway between the present-day communities of New Iberia and Loreauville. There de Vaugine constructed a comfortable residence by frontier standards. Standing on sturdy horizontal beams, the house was "divided into three rooms enclosed by galleries on two sides with two storehouses made of posts in the ground, covered with straw, one of which serves as a drying room with . . . court and garden enclosed by pickets. . . ."[32]

De Vaugine built this house for himself, his wife, Pélagie Petit de Livilliers, and their several children. But Pélagie died prematurely in 1772, leaving de Vaugine to "make a separation of the community of property belonging to him and . . . his [late] wife, and to ascertain the portion belonging to each of the heirs." For this purpose de Vaugine asked local commandant Fuselier de la Claire to draw up a detailed inventory of the estate. Although de Vaugine was undoubtedly more affluent than most of his neighbors, the inventory provides a rare glimpse into the material culture of a prosperous eighteenth-century indigo plantation on the banks of the Teche.[33]

For example, the inventory shows that the dwelling held sideboards, armchairs, chests, tables (including one for playing the popular card game quadrille), andirons, and a four-poster bed of walnut with yellow satin curtains and feet carved like deer hooves. The house contained an array of raw cloth, including ells of colete, batiste, royale, and coutil, along with pieces of bretagne and moiré ribbon and handkerchiefs of *indes rouge* (red indigo). The kitchen included ordinary culinary items—forks, spoons, knives, pots, pans, kettles, teapots, saucepans, spits—but had nearly 140 plates, including 42 made of the glazed pottery called faïence, along with 24 faïence coffee and tea cups, 24 goblets, 96 butcher's knives, and 180 napkins. De Vaugine also had on hand a hundred pounds each of sugar and coffee, and "two fine muskets" with fifty pounds of gunpowder and eighty pounds of lead shot. Among the most personal items recorded in the inventory were "five rings mounted in carnelian, marcasite, and topaz in gold."[34]

Outside the de Vaugine residence stood the vats for making indigo (each with a chain pump for drawing water) and a horse mill for thrashing the indigo plants. Among the tools on hand to maintain the place were shovels, adzes, planes, hammers, saws, chisels, augers, clamps, vices, grindstones, plows, and axes (including two used "to kill, cut up, and skin animals")—and even a still for making alcohol. Around the plantation could be found many horses, cows, oxen, pigs, and mules. For transportation de Vaugine had three ox carts, a boat of unknown size—a necessity for living on the Teche—and at least a couple *traîneaux*, drag sleds pulled by horses or oxen across solid ground or mud. (This now-extinct type of vehicle is immortalized in the popular Cajun song "*Ils ont volé mon traîneau*" [They Stole My Sled].)[35]

Listed in the inventory, de Vaugine's debts reveal a personal financial network that connected his seemingly isolated plantation to the outside world, even to Europe. He owed creditors in New Orleans, the Illinois country, Haiti, and across the Atlantic in La Rochelle, Strasbourg, and Paris (including one hundred twenty piastres due "Mr. Pinard, tailor of Paris").[36]

And then there were the slaves—thirty-five in all, ranging in age from five months to fifty (although two were listed merely as "old" and "decrepit"). Most were aged forty and under—the average age being about twenty-five—and their number included two married couples, four unmarried mothers with nine children between them, and an unmarried father with a six-year-old son. They bore names like Charlot, Samson, Sans Souci (meaning "Without Worry"), La Roze, Sauvier, Izabelle, Bacchus, Fanchonette, Marthon, and Gros Louis ("Big Louis"). De Vaugine specifically identified three of his slaves as "Creoles" (Louisiana natives), suggesting the remainder had been born elsewhere, perhaps Africa or the Caribbean. One slave, Louis, bore the additional designation "Congo," suggesting his place of origin on the west coast of Africa.[37]

Like slave owners in general, de Vaugine used violence, or at least the threat of violence, to instill obedience. Three items in the plantation inventory serve as proof of this practice: "*deux paires d'enferges pour negres* [two pairs of iron shackles for Negroes] . . . *deux anneaux pour negres malfaiteurs* [two iron rings for Negro malefactors] . . . *une barre a prisonnier*" [iron shackles connected by a solid metal bar, called in English *bilboes*]. Despite such barbaric tools of punishment, or perhaps because of them, two of de Vaugine's slaves, Guillaume and Jasmin, had run off months earlier to New Orleans, where they disappeared among the city's burgeoning black population.[38]

Besides de Vaugine, the Teche supported at least a handful of other commercial indigo planters. Judge Jehu Wilkinson of Franklin, a Pennsylvania native who arrived on the Teche in 1810, recalled in 1847 that "Sorel [*sic*] and others . . . had previously made indigo" before switching to cotton and sugarcane. Those other early indigo growers on the Teche included Jean-François Ledée, who in 1778 wrote to Alexandre DeClouet, commandant of both the Poste des Attakapas and Poste des Opelousas, of "*mon indigoterie*" (my indigo works). Unlike de Vaugine, however, Ledée was primarily a rancher, owning 450 head of cattle (compared to de Vaugine's meager 26 head).[39]

Several devastating factors—including strong competition abroad, repeated pest infestations, and rampant plant disease—destroyed indigo as a major Louisiana industry by the early 1800s. Regardless, one Teche Country resident recalled that "many indigo vats were to be seen along the Teche" as late as the 1830s. These possibly belonged to the Acadians, who continued to engage in small-scale indigo farming for household use. Skilled at weaving homespun cloth, Acadian women used indigo to add color to their otherwise natural brown or yellow fabric.[40]

Ultimately settling only about five miles downstream from de Vaugine's plantation, another ethnic group, the Spanish, joined the Acadian and French farmers along the bayou. In late 1778 governor Gálvez commissioned one of his soldiers, Francisco Bouligny, to establish an outpost on the Teche. Bouligny recruited colonists for the endeavor from the Spanish port of Málaga, located on the Mediterranean coast about eighty-five miles northeast of Gibraltar. Born not too far from Málaga in the port of Alicante, Bouligny had arrived in Louisiana a few years earlier under Alejandro "Bloody" O'Reilly (so-named because of his brutal suppression of the Rebellion of 1768, when Louisiana colonists overthrew their first Spanish governor, Antonio de Ulloa). Remaining in the colony, Bouligny rose to the rank of lieutenant colonel in the Fixed Regiment of Louisiana and developed a genuine interest in the impoverished Spanish possession. He suggested to Gálvez the founding of a settlement deep in the interior to expand agricultural output, boost the colony's population, and defend the territory against British intrusion.[41]

Approving the plan, Gálvez nonetheless rejected Bouligny's proposed site for the settlement, Ouachita (in present-day northeast Louisiana). That location, asserted the governor, sat too far from New Orleans, too close to hostile Native Americans, and would inflate the cost of the venture. Bayou Teche, however, lay closer to the capital, had a friendly tribe on its banks, and ran through much unclaimed land said to be extremely fertile. Additionally, the French, French Creole, Acadian, and other settlers already on the Teche—including an increasing number of African slaves and *gens de couleur libre* (free persons of color)—could no doubt help Bouligny's colonists survive.[42]

Bouligny and an advance party of twenty Málagueños left New Orleans for the Teche in late January 1779. With them they took rowers, soldiers, artisans, and over thirty slaves. They followed the approximate route that Beausoleil and his fellow Acadian exiles took a few years earlier: up the Mississippi, then west across the Atchafalaya swamp. Burdened by tools, weapons, and provisions, the expedition slogged its way through the foreboding wetlands in a journey described by one soldier as "fifteen days of Purgatory."[43]

"[A]bout the 11th of this month," Bouligny reported to Gálvez around mid-February, "we entered the Rio Teche." There someone awaited them: Commandant DeClouet. A former French officer, DeClouet had allegedly fled Paris after an amorous misadventure with a sister of the duc de Choiseul, the French king's close advisor. Once in Louisiana, the Spanish had appointed him commandant of both colonial *postes* along the Teche. In this capacity DeClouet met Bouligny to discuss settlement sites on the bayou.[44]

Less than a week later Bouligny wrote Gálvez, "I decided to establish the town near the Chitimacha[,] whose leader is Soulier Rouge"—Red Shoe, ceremonial title of chief Kamin Nu Tuh. Bouligny continued, "I intend to assign each settler six arpents of land fronting the Teche on the right bank going up [i.e., the east bank] for cultivation. I will also grant them six on the left bank where I will found the town and where I will leave the land in common for grazing. . . ."[45]

Two *gens de couleur libre* resided on the site chosen by Bouligny. One moved after Bouligny promised him land farther down the Teche; the other DeClouet simply banished. Bouligny paid the Chitimacha fifteen pesos for the land and its improvements (two cabins and a fence), adding "about a hundred pesos worth of presents"—perhaps some of the thirty pounds of glass beads carried on the expedition.[46]

When it came to laying out the village, Bouligny assigned the most grueling tasks to the enslaved blacks. In less than two months these slaves plowed twelve to fifteen arpents for seeding, built two palmetto warehouses for storing communal provisions, constructed a large coal-fired oven, hewed fifty eighteen-foot timbers felled in a nearby cypress grove, cut logs for a bridge to span the 240-foot width of the Teche, and erected palmetto huts for the Málagueños, soldiers, warehouse guard, and blacksmith, as well as for two jacks-of-all-trades who helped with construction, surveying, and cartography. The slaves also fenced a public corral for the village livestock. Purchased from DeClouet, these animals included thirty-two teams of wild and domesticated oxen, six horses, "and many pigs and chickens."[47]

Bouligny christened the village Nueva Iberia, or New Iberia. There the Málagueños sowed hemp, flax, wheat, and barley. Although the hemp and flax seeds failed to germinate, the wheat and barley thrived even in the subtropical climate. Bouligny soon boasted to Gálvez of the settlement's "beautiful location and fertile lands," adding, "[I]t would be difficult to find a place that includes so many advantages."[48]

The next month, however, rain fell incessantly. Every day the Teche rose over a foot. The turbid water soon overran its banks and inundated the Málagueños' village. "Flood unlike any previously known," recorded Bouligny. "The houses or huts the families had erected on their own land were under six to eight feet of water. . . ." The deluge extended northward, he observed, swamping "the establishments of Mr. Flamand [Grevemberg], Fuselier, Le De[e], Masse, and many others. . . ." Even the Chitimacha found their huts "covered . . . almost entirely."[49]

"[A]lmost all communications on the Teche are interrupted," lamented Bouligny.[50]

He responded to the catastrophe by moving Nueva Iberia about twenty miles upstream to the foot of the Fausse Pointe oxbow, at a place called Petite Fausse Pointe. He chose a spot on the west bank near a *portage* (land over which boats could be carried from one waterway to another) and not too far from a lake "filled with fish and game"—either the interconnected lakes to the northeast and east known today as Lake Dauterive and Lake Fausse Pointe or, more likely, Spanish Lake to the northwest, known in Bouligny's time as *Lac Flamand* (an allusion to the Grevembergs) and *Lac Tasse* (*tasse* meaning *cup*, from the lake's "round-as-a-cup" appearance). The site also attracted Bouligny because nearby a small coastal inlet—possibly a good harbor for large seagoing vessels—opened through Vermilion Bay to the Gulf of Mexico. Called *Petite Anse* (literally, "Little Cove"), this lagoon eventually gave its name to an adjoining bayou and to the looming salt dome now known as Avery Island.[51]

At Nueva Iberia's new site, Bouligny ordered the construction of two sixty- to seventy-foot sheds as temporary shelters for the Málagueños, slaves, soldiers, and other settlers. This time, however, he hired skilled Acadian exiles to erect the permanent dwellings for his colonists. To preclude the possibility of flooding, he designed these new homes to stand nine feet off the ground. After overseeing the resettlement, Bouligny financed exploration of the Teche below Nueva Iberia, sending a survey team down the bayou to Atchafalaya Bay, then west to Petite Anse Island, and up Bayou Petite Anse and the coastal prairie to Nueva Iberia. Bouligny's expedition, however, came only at the end of what might be considered the golden age of Teche exploration. That golden age began a few years earlier, in winter 1769.[52]

2

Exploring the Bayou

Because the earliest known reference to the Teche dates from the Spanish rule of Louisiana, it is unclear if earlier French colonial officials knew the bayou existed. The settlement, however, of a handful of colonists along the Teche during the French period—including André Masse, Gabriel Fuselier de la Claire, Jean-Baptiste Grevemberg, and Jean-François Ledée—suggests French authorities were aware of the waterway, even if they might not have known its exact location or even its name. Regardless, during their six-decade rule the French never bothered to formally explore south Louisiana's interior beyond the banks of the Mississippi. As a result, official investigation of this hinterland, including the Teche Country, occurred only after Spain took charge of the colony. This meant that numerous pioneers, including the roughly two hundred Acadians under Beausoleil, had already settled along the Teche by the time administrators sent colonial agents to explore the region.[1]

After France ceded Louisiana to Spain in 1762, seven years lapsed before the colony's second Spanish governor, an Irishman by birth named Alexandro O'Reilly, dispatched the Kelly-Nugent expedition to reconnoiter the Teche region. Like their commander, Juan Kelly and Eduardo Nugent were Irish-born soldiers in service to the Spanish king. O'Reilly ordered the two officers to journey from New Orleans through the Attakapas and Opelousas districts to the presidio (fortified village) of Los Adaes, capitol of Spanish Texas near Natchitoches. The purposes of the expedition were to extract oaths of allegiance from colonists; to count all inhabitants, free and enslaved, along with every horse, mule, donkey, cow, sheep, goat, pig, ox, and cart; to report on the condition of roads and trails in case the Spanish military should need them for deployment; to receive petitions and complaints; and to collect the names of local troublemakers. O'Reilly also instructed Kelly and Nugent to assess the quality of the land and the types of produce grown in the places they visited.[2]

With horses, slaves, and other Spanish soldiers in tow, Kelly and Nugent departed New Orleans in mid-November, following the Mississippi River northwest to Bayou Plaquemine. There they first encountered the winter rains that would plague their mission. And there, on the western bank of the Mississippi near the edge of the massive Atchafalaya swamp, half the expedition became lost in "the great downpour" and "wandered aimlessly in the woods." After local Native Americans found the missing party, the expedition regrouped and headed into the Atchafalaya, reaching the Attakapas region after a three-day journey.

Although Kelly and Nugent's journal never mentions the Teche by name, the expedition clearly followed the bayou upstream for at least fifty miles of the trek toward Los Adaes. Reaching the home of "Flamand" (one of the Grevembergs), the explorers traced the course of the Teche on foot to the residence of Gabriel Fuselier de la Claire, located near the junction of bayous Teche and Fuselier (near present-day Arnaudville). From there they continued upstream to the homestead of Jacques Courtableau near the headwaters of the Teche (around present-day Port Barre). The party arrived, however, only after once more becoming lost in the wilderness. Again, Native Americans, along with other locals, "came to our assistance," recorded Kelly and Nugent, "attracted by the continuous gunshots which we fired in the woods."

The journey up the Teche had by then become an ordeal. Rain fell incessantly, filling the bayou, turning trails to muck, and miring the horses. Settlers urged the expedition to turn back, but it pressed forward, felling trees to make bridges across the turbulent water. Kelly and Nugent tried to secure local Native American guides, but "not a single Indian would undertake [the journey] with us," the soldiers chronicled, "stating that it would probably mean death. . . ."

"[T]here were such troubles and uncountable hardships during this trip," they complained, "that we, having now seen the dangers, considered it almost impracticable to continue the trip this way."

Despite these adversities, the expedition made perceptive observations while advancing up the swollen Teche. "Atacapas and Opelusas [*sic*] are two separate districts," Kelly and Nugent recorded, "divided by a small bayou [Bayou Fuselier] which flows by Fuselier's [homestead]. However, they can be considered as one, wholly alike in quality of land, products, and livestock." The terrain, they remarked, consisted of "spacious prairies covered with admirable grazing of very high and slender grass which is free from thistle and thorn. . . ." The soil yielded corn, rice, and sweet potatoes, and many of the inhabitants specialized in cattle ranching. "The care of the cattle keeps the natives busy," Kelly and Nugent observed, "though it does not necessarily mean much work. . . ." Still, they noted, dashing a later stereotype of the lazy Acadian, "The inhabitants are not indolent and among them there are some industrious Acadians . . . [who] live in great tranquility and accord. . . ."

Before departing the Teche Country for Natchitoches, Kelly and Nugent counted between the Attakapas and Opelousas districts a total of 363 whites, 148 slaves, and approximately 6,200 farm animals, 3,700 of which were cattle. These figures reflect the character of frontier life along Bayou Teche in the late eighteenth century.[3]

A few years later an Anglo explorer and cartographer named Thomas Hutchins compiled his knowledge of the Teche—not for Spain, but for its chief enemy, England. Hutchins, it seems, was "Most certainly a spy," historian Joseph Tregle maintains. Born in New Jersey in 1730, Hutchins became a British soldier and military engineer. He helped drive the French from Fort Duquesne (present-day Pittsburgh), used his surveying skills to map the Ohio and Illinois countries, and rose to the rank of captain. In 1772 he strengthened fortifications in British-held West Florida before receiving orders to secretly reconnoiter Lake Borgne, Lake Maurepas, Lake Ponchartrain, and New Orleans. Hutchins examined strategic approaches to the city and evaluated the strength of its defenses—intelligence that could be used to assault New Orleans should war break out again with Spain. (Four decades later British commanders might well have consulted Hutchins's work in the run-up to the Battle of New Orleans.)[4]

Hutchins remained in service to the crown during the early years of the American Revolution, bolting to the rebels—with assistance from Benjamin Franklin—only after his imprisonment in London for "high treason." (Prosecutors released him, finding he had merely passed coded financial information, not state secrets, to an American entrepreneur.) Returning to America as a newly converted revolutionary, Hutchins secured a congressional appointment as "Geographer to the United States," a post he retained until his death in 1789.[5]

While reconnoitering the Gulf Coast in 1772 and 1773, Hutchins collected information about Bayou Teche (which he called "the Tage River"). It is unclear if the British spy explored the bayou for himself or relied on secondhand intelligence to compile his report. Tregle, however, believed that Hutchins ascended the bayou in person. As the historian observed, superiors ordered Hutchins to "explore Lakes Pontchartrain and Maurepas, the Amite River, Bayou Manchac, *the Teche*, and the course of the great Mississippi" (my italics). Tregle additionally noted:

> It is in his account of the regions west of the Mississippi, along Bayou Teche, or the Tage River, as he called it, and down the Atchafalaya, that Hutchins becomes once again the true pioneer, reporting upon a region which those before him had completely neglected. Charlevoix, Bossu, Le Page du Pratz, La Harpe, Dumont—none of these earlier writers give any notice at all to this area which Hutchins immediately recognized as of great consequence in the eventual growth of the territory. For this reason his pages on the Teche and the surrounding countryside are among the most significant in his work and give it a place of particular importance in the literature on Louisiana.[6]

But exploring the Teche in person would have been difficult for Hutchins. Unlike other waterways he investigated along the Gulf Coast, the Teche did not adjoin British territory. As such, he could not have claimed right of navigation, as he had done elsewhere, or masqueraded as a harmless surveyor and engineer. Still, English-speakers on the Teche were not unheard of—Thomas Berwick and William Henderson, for example, settled on the bayou by end of the 1770s—and Hutchins could have used other pretexts to bluff his way up the bayou.[7]

Regardless, Hutchins clearly relied at some point on others for information about the Teche. His revised report, for instance, mentions "la Nouvelle Iberie," a place founded six or seven years after his mission to the Gulf Coast. Perhaps his informant was the same person he credited with supplying details about the Apelousa River (Bayou Courtableau), a certain "Dr. Lorimer" (probably Dr. John Lorimer, a British military physician in West Florida), who in turn obtained his information from "a Frenchman" of unknown identity.[8]

Hutchins began his description of the Teche at its mouth—actually the mouth of the present-day Lower Atchafalaya, which connects to the Teche some ten miles upstream—providing distances from one landmark to another in Spanish leagues (1 league equaling 2.63 miles). On the Lower Atchafalaya he noted the location of Prairie de Jacko (or Jacob, as he also spelled it); then, entering the actual Teche, the abandoned Native American village he called Mingo Luoac; the homestead of "Mons. Mass[e]" near present-day Baldwin; and the inhabited Chitimacha village of Selieu Rouge. He next identified the homestead of "Mons. Sorrel," on the Teche just southeast of present-day Jeanerette, "From whence," he observed, "to the town of la Nouvelle Iberie, on the same side [of the bayou], it is six leagues." He added, "The whole of this distance is tolerably well settled."

Having guided his reader this far up the bayou, Hutchins paused to make a few general observations. He noted, for example: "The river Tage [Teche], is in general better than 100 yards wide, with a gentle current, and a small ebb and flow of about 8 or 10 inches. It narrows as you ascend it, where in some places, it is not 50 yards over. Vessels drawing from 7 to 8 feet water may go from the sea to this town [New Iberia] without any obstructions."

Continuing up the Teche, Hutchins identified "la Force Point" (Fausse Point), followed by a rare contemporary reference to "la Shute," a small waterfall "of about 10 feet [high]" spilling from a small tributary into the Teche—a landmark that seems almost as misplaced on the Teche as an iceberg. (The tributary, Bayou La Chute, still exists, but the waterfall has long since disappeared; still, its memory is preserved

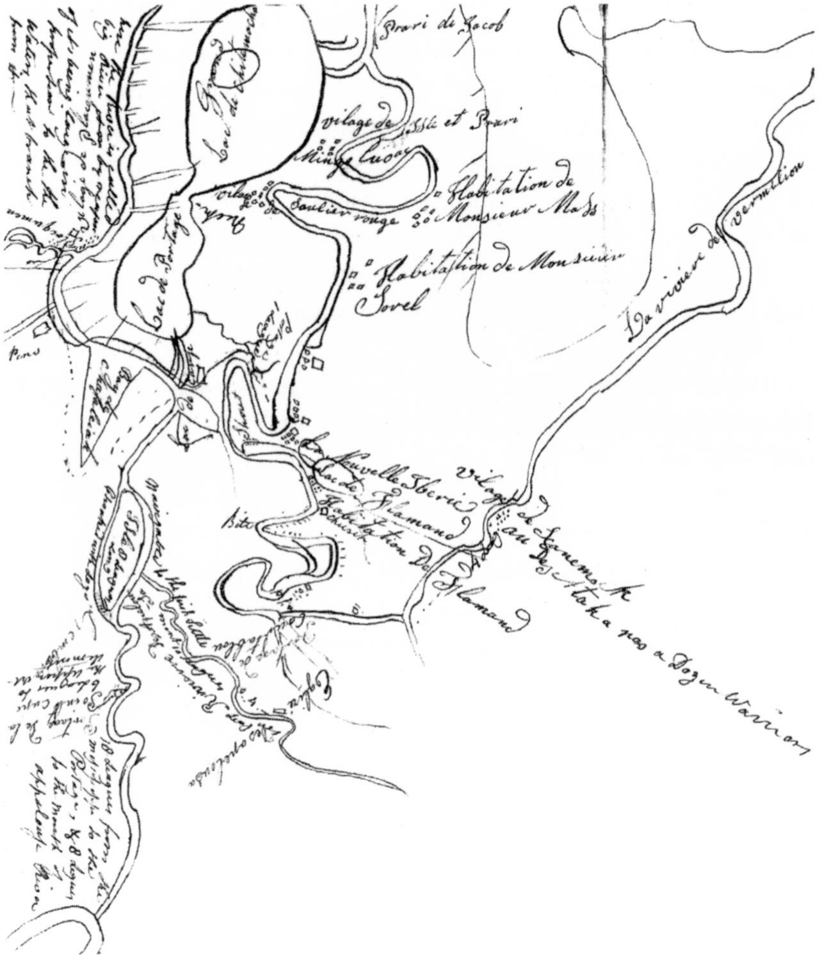

Thomas Hutchins's hand-drawn map of the Teche (or "Tage," as he called it) is one of the earliest maps of the bayou, dating from between 1779 and 1784. The Teche is the waterway in the middle; the Atchafalaya River is at far left and the Vermilion River is at far right. (Hutchins drew the map upside down.) Source: Historical Society of Pennsylvania, Philadelphia.

in the very name of the bayou: *chute* in French means "waterfall.") Beyond this curiosity lay the homestead of "Mr. Flemming" (Grevemberg dit Flamand) and, farther up the Teche, "the church Desata cappau"—Hutchins's phonetic mangling of "des Attakapas." The structure sat on or near the site of today's St. Martin de Tours Catholic Church in St. Martinville.

Only two leagues past the church, according to Hutchins, the Teche jutted suddenly to the east before turning back on itself and resuming its previous course to the north. This oxbow, he explained, was known as "the bottom of the bite" (that is, *bight*, meaning a curve or bend). Although Hutchins exaggerated this feature on his hand-drawn map of the Teche, he referred no doubt to the bend at present-day Parks, known at the time to local settlers as *La Pointe de Repos*. "From thence to the point settlement of Acadians is one league," he went on, referring to exiles who had settled just above the bend, though some had also settled below it.

Upstream he identified the plantations of "Mons. l'Deé" (Ledée) and "Mons. Fuzelliere" (Fuselier de la Claire)," where "Fuzellier's fork, or branch [Bayou Fuselier], is just below his house, and divides the districts of Attacappau [Attakapas] and Appalouse [Opelousas]." He pointed out only two features north of the Teche's junction with Bayou Fuselier: "the Prairie de Mons. Man" and "Mons. Man's plantation," both referring to François Manne, whose two-thousand-acre grant stood on the west bank of the Teche near the site of present-day Leonville.

Beyond that point, recorded Hutchins, the Teche "divides itself into little brooks, and loses itself in rich and extensive savannahs." Hutchins's hand-drawn map likewise shows the Teche dying out south of the Rivière des Apelousa (Bayou Courtableau). This could be considered an accurate description of the upper Teche during low-water months. In that season the bayou's channel ran empty above the site of present-day Arnaudville, and yet might have seemed to "divide itself into little brooks," namely, Bayou Little Teche, Bayou del Puente (now often spelled Puent), and Bayou Toulouse, all minor waterways connecting to the upper Teche.[9]

Another explorer, the Spanish mariner José Antonio de Evia, also ascended the Teche during this period. Born to a master mariner in La Graña on the Bay of Biscay, Evia attended a royal naval academy and became a pilot, sailing the Gulf of Mexico between the Spanish ports of Vera Cruz, Havana, Mobile, and New Orleans. He is credited with the pursuit and seizure of at least two British vessels, as well as the capture of British-born adventurer William Augustus Bowles, whose crimes included the establishment of a short-lived Native American nation in Spanish Florida, not to mention an ineffectual declaration of war against Spain.[10]

In 1783 Gálvez ordered Evia to explore the Gulf Coast from Florida to the port of Tampico, Mexico. After a false start that year, he set sail in 1785 with two *goletas* (schooners) named the *Grande* and *Chica Besaña*, reaching the mouth of the Lower Atchafalaya in mid-June. He called the waterway "the Rio Chafalayá, or Teche," for like many at the time Evia regarded the Lower Atchafalaya and the Teche as synonymous. On spotting the first dwelling four leagues [about eleven miles] up the Lower Atchafalaya, Evia dropped anchor to go farther up the Lower Atchafalaya and Bayou Teche in a smaller vessel. "I took a *piragua* [pirogue] and with three sailors," he recorded in his journal, "I went up Atacapas [*sic*] to the home of Commander Dn. [*Don*, an honorific] Alexandro DeClouet, which is 35 leagues [about 92 miles] from said place." This indeed equaled the distance to DeClouet's home on the bayou just below present-day Cecilia, if one made a portage at New Iberia to avoid rounding the Fausse Pointe oxbow.[11]

"The [first] four leagues that there are from the entrance to the first settlements of this river [the Lower Atchafalaya and then the Teche proper]," he observed, "is land that is swampy and impassable." Beyond that, however, he found the land "high, fertile and pleasant" with an "abundance of all species of livestock and game." Moreover, the land "produces much fruit, indigo and tobacco, all of which is carried in pirogues to New Orleans." As for settlers, Evia found the banks of the Teche "all populated . . . with Spanish, French, Germans, and Americans, and many peaceful Indians, called the Opelousas and Attakapas. . . ." Actually, Evia most likely encountered the Chitimacha, whose village he would have passed at the site of present-day Charenton. Oddly, Evia made no reference to Nueva Iberia or the church at the future site of St. Martinville, both of which he certainly must have seen during his journey up the Teche.[12]

On finding DeClouet, Evia explained to the commandant "the orders with which I found myself, and the reasons for my arrival": to hire boats and soldiers for the purpose of exploring the Louisiana and Texas coasts as far as "*la Bahia de Sn. Bernardo*" (San Bernardo Bay), near the site of present-day Galveston. In response DeClouet recommended that the Spanish captain meet with the area's most experienced pilots, who could provide him with navigational advice as far as "*el Rio Savina*" (the Sabine River, present-day border of Louisiana and Texas). When Evia met these pilots, they warned him that oyster banks to the west would block his deep-drafted schooners. They recommended he leave his schooners on the Lower Atchafalaya and instead use *piraguas* and *verchas* (that is, *berchas*, flat-bottomed barges propelled by oars and perhaps sails) to explore the Texas coast.[13]

Evia took their advice, but armed these smaller vessels, fearing attack by "*los Yndios Carankaguases*," the Karankawa tribe of southeast Texas—reputedly the killers of Andry, the respected French engineer who a few years earlier had escorted the Acadian exiles under Beausoleil to their new home on the Teche. As Evia recorded, "I leased two barges and two pirogues, and added twenty militiamen of Atacapàs and Opelusas . . . which well-armed could defend against the . . . Indians, if we are attacked." Those militiamen consisted entirely of "*Negros, y Mulatos*," all volunteers led by the commandant's son, Louis DeClouet. The younger DeClouet, noted Evia, possessed "the special ability to speak the language of those Indians." Finally, Evia borrowed the elder DeClouet's best pilot for the voyage.[14]

Taking the barges, pirogues, and militiamen, Evia and his sailors descended the Teche and Lower Atchafalaya to the Gulf. They then turned west to follow the coast to San Bernardo Bay and the Rio de los Orkoquisas, or Rio Trinidad, in the territory of the feared Karankawa. The expedition saw nothing of these natives, however, except footprints and distant bonfires. After charting the region, Evia returned to the Teche in early August, gave DeClouet back his vessels and soldiers so as "not to cause more expense," then sailed his schooners to New Orleans before again steering toward the west to explore the coast to Tampico.[15]

But it is the expedition funded by Francisco Bouligny in summer 1779 that the historical record documents in greatest detail. Shortly after Bouligny established Nueva Iberia at its permanent site on the Teche, he commissioned two non-Spanish settlers, local landowner Jean-Baptiste Grevemberg and royal surveyor François Gonsoulin, to explore the Teche from the village to the bayou's mouth ("*la salida del Theis*," as Bouligny called it). Once they passed the mouth and, after

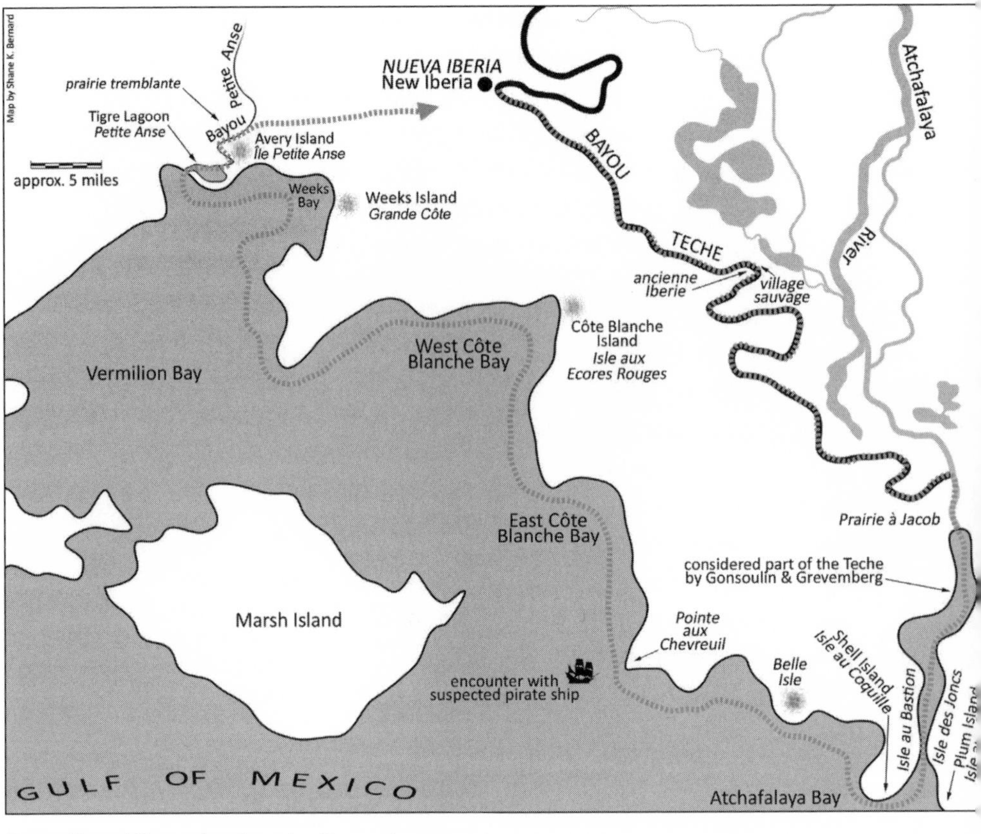

Gonsoulin and Grevemberg's exploration route, 1779.

more rowing, arrived at the Gulf of Mexico, the expedition would follow the coast west to the lagoon called Petite Anse ("*Pequeña Anza*," in Bouligny's tongue), then paddle up the bayou of the same name until the water ran out and finally walk the last few miles back to Nueva Iberia.[16]

Gonsoulin and Grevemberg would measure the bayou as they traveled it, logging the distance in *toises*—the *toise* being an archaic French unit of measurement equaling six feet. Likewise, they would sound the bayou's depth in *pieds* (feet) and, in deeper spots, *brasses* (fathoms). And while both men, with their slaves, worked tirelessly to survey the route, it was Gonsoulin who not only wrote down the measurements but kept an impressionistic, and intriguingly readable, journal of the expedition.[17]

Hailing from Marseilles, Gonsoulin had studied piloting, served thirty years in the French navy, and captained a merchant ship prior to coming to Louisiana. He settled in the colony, Gonsoulin noted in the journal, "to establish here my residence, having bought a farm two leagues [about five miles] from Nouvelle Iberia, in order to live and die under his Catholic majesty [the Spanish king], in the capacity of a farmer and planter of this beautiful land. . . ." Bouligny praised Gonsoulin to Gálvez, noting the Frenchman's "reputation for being a very skilled pilot" and his "genius and passion" for discovery.[18]

Drawing on his navigational skills, Gonsoulin stood in Nueva Iberia's "*place d'armes*" on the expedition's first day, June 18, 1799, and confirmed the village's latitude. Grevemberg and he then departed down the Teche with six African slaves in Bouligny's personal *falua* (a small open boat propelled by oars or one or two sails—the latter of which would have been useless on the bayou). The explorers took with them a month's provisions, including fishing line, sailing canvas, liquor, and several yards of fabric as *mosquiteros* (mosquito netting). They also carried flint, gunpowder, birdshot, and bullets for their firearms, as well as a *pedrero*, a small cannon they would fire daily to convey their location to those back at Nueva Iberia. For this same purpose Bouligny ordered Gonsoulin and Grevemberg to set fire nightly to the tall, reedy bulrushes that grew along the bayou and coastline. The glow on the horizon would relay their progress.[19]

With the crew of slaves working the oars, the falua glided down the bayou, passing, as Gonsoulin recorded, the "*village sauvage*" (Chitimacha village), "*ancienne Iberie*" (the original site of Nueva Iberia), and, finally, "*les dernières établissements*," the last pioneer dwellings on the waterway. They floated past the "*prairies tremblantes*," where water so permeated the marshy earth that the ground trembled when disturbed even by footsteps. The expedition soon reached what is today considered the end of the Teche (at the site of present-day Patterson) and entered a wider body of water more like a lake. Modern maps identify this stretch as the Lower Atchafalaya River. Gonsoulin, however, regarded it a natural continuation of the Teche. Indeed, he bristled at those like Bouligny who called the Lower Atchafalaya "*el Rio Grande*" or in French "*Grand Rivière*." They spoke "inappropriately," asserted

Gonsoulin in his diary. On the contrary, the great river, he insisted, was "always the Teche."[20]

"The Teche is full of all species of the sea," boasted Gonsoulin as the slaves now rowed south toward the Gulf of Mexico. The falua passed a flat grasslands, which Gonsoulin identified as "*la Prairie à Jacob*," or Jacob's Prairie. (This is the same stretch Hutchins had called the Prairie de Jacko and the Prairie de Jacob.) The explorers soon spotted two islands in the wide, turbulent river. One of these they named *Île des Joncs* (Bulrush Island) and the other, a "very small but high" point, they dubbed *Île au Bastion* (Bastion Island). They gave it this name because, as Gonsoulin suggested, the place "might be good for a fort." (Evia would use their name for this island when he ascended the waterway a few years later.)

Beyond this point the river wrapped around another, much larger island about five miles in circumference. There the group debarked to camp and explore. Gonsoulin found the place rich in soil, teeming with bear and deer, and forested in "*chêne vert*" (live oak). Instantly enamored of the place, he waxed poetic in his diary: "The grandeur, the beauty of the trees, the earth, covered in hills and valleys, its vegetation on high ground, the view of the sea, edged by a piece of land rather high in reeds, which will prevent it more and more from erosion by the waves—that say I, by the pleasure that we felt by looking at it, made us call it *Belle Isle*."

And so the place is called today. Other places still bearing names given them by this 1779 expedition include *Pointe aux Chevreuil* (Deer Point), known even now by the original French; as well as *Île aux Prunier* and *Île au Coquille*, both known today by the English translations Plum Island and Shell Island. Other places the expedition named—*Île Ronde* (Round Island, not to be confused with the present-day place of the same name on Bayou Chêne), *Île au Lattanier* (Palmetto Island, likewise not to be confused with the present-day state park in Vermilion Parish), *Île aux Goëlan* (Seagull Island)—elude identification and perhaps now have entirely different modern names. Or perhaps they no longer exist, destroyed by over two centuries of storms, erosion, subsidence, and dredging. However, the features the explorers called *Les Quartres Soeurs* (The Four Sisters), never identified in their journals, might have been their collective name for the four massive salt domes visible from offshore: Belle Isle, Côte Blanche, Weeks Island, and Avery Island. (The fifth major salt dome in the region, Jefferson Island, lay farther island and could not be seen from the Gulf.) One of these salt domes, named *Île aux Ecores Rouges* (Red Bluffs Island) by the expedition, today bears a French moniker denoting a different color, *Côte Blanche* (White Coast or White Hill—*Côte* can mean "coast" or "hill," and because the object in question looks like a hill and sits on a coast it is impossible to know which meaning the region's early French speakers had in mind). As the falua shadowed the coastline, however, Gonsoulin and Grevemberg came to realize that many of the "islands" they had named were not islands at all; rather, they were fragments of the mainland sliced through by bayous and wetlands, and only appeared insular.[21]

Everywhere the explorers saw evidence of the region's copious natural resources. For example, they discovered an abundance of sassafras trees, from whose fragrant leaves the colony's Native Americans made the *filé* used to season and thicken that essential south Louisiana dish, gumbo (known to exist in the colony as early as 1764). They found woodlands rich in "*bois de construction*" (timber suitable for building) and in one inlet they noted an unidentifiable plant that produced "a gum which . . . made on paper a fine green [color] of much beauty." Intrigued by a variety of fragrant yellow flower, Gonsoulin suggested its propagation back at Nueva Iberia, if only for the beauty and agreeable scent. Elsewhere the expedition found plum, palmetto—even prickly pear cactus, which grew on the colony's sandy *chênières* (coastal oak-grove ridges) and elsewhere.[22]

Offshore the explorers found the bottoms replete with oyster reefs, a potential source of nourishment for the colonists, and one that local Native Americans had tapped for over a millennium. "The length of the coast," Gonsoulin recorded, "is teeming with deer," and bear, too, he observed. He spotted feral livestock (*les marrons*) grazing on the coast, fugitives from a shipwreck or some unlucky colonist's *vacherie*.

"If God descended to Earth," he enthused, "He would choose for his stay this beautiful country."

The journey was not without a sense of danger. Near Pointe aux Chevreuil the explorers steered a course into deeper water. A tall ship appeared on the horizon, turned toward the falua under full sail, and fired a distress call with a cannon. As the larger vessel drew near, the explorers discerned it was (according to Bouligny's retelling of the event to Gálvez) a "*bergantin*," or brigantine, a small, fast two-masted ship. At its sight Gonsoulin felt "*la peûr d'un corsaire*," as he wrote—the fear of a pirate. And not without reason: Gonsoulin, the old sailor, found it suspicious that a larger, seagoing vessel should in distress hail his tiny open boat for assistance; to him it seemed a ruse. Moreover, a seasoned mariner like Gonsoulin would have known that pirates preferred to sail in brigantines. (Indeed, the word *brigantine* derives from same root as *brigand*, meaning bandit or robber—an association that reflected the ship's popularity with pirates.)[23]

Gonsoulin, Grevemberg, and their slaves made rapidly for shore, ducking into an inlet until the ominous ship departed. Then they continued on their way, the falua at last reaching the cove known as Petite Anse. Almost four miles north of this lagoon sat a compact range of wooded hills rising over 160 feet above the encircling salt marsh. A 2,200-acre salt dome, the prominence had been pushed up over eons from deep inside the Earth. Its natural saline springs had attracted animals for over ten thousand years, including (as shown by the fossil record) mastodons, mammoths, giant sloths, bison, and wild horses. The geological oddity also attracted early Native Americans, who boiled its briny spring water in clay pans to extract the salt. Countless sherds of pottery, mostly undecorated but some bearing lines and swirls and even ancient fingerprints, still litter the place as evidence of this prehistoric activity.

This salt dome would be known by many names: during the late colonial period the Spanish called it *Isla Cuarin*, while the French named it *Côte de Coiron*—Cuarin Island and (again, because of the indistinct meaning of the word *côte* in this context) Coiron Coast or Coiron Hill, all referring to an early claimant, Antoine Coiron. In the nineteenth century, however, it would be called Petite Anse Island and, later, Avery Island. The place bore no name in Gonsoulin's day; he merely called it "*la pretendü isle que nous voyons de la mér*" (the so-called island that we saw from the sea).[24]

There the falua turned north up Bayou Petite Anse, which appears in late-eighteenth-century documents as *Bayou de Petite Côte*. Gonsoulin, Grevemberg, and their rowers followed the bayou along another *prairie tremblante* north of Avery Island. When the bayou ran too shallow, the explorers grounded the falua and completed the rest of the journey back to Nueva Iberia and the Teche on foot. Pleased by the expedition, Bouligny went on to commission Grevemberg to explore a water route through Barataria Bay to New Orleans. Gonsoulin, for his efforts, received a Spanish concession of about 400 acres of high land and about 2,000 acres of adjacent marsh—all at that dreamy place along the sea, Belle Isle.[25]

3

From a Colonial to an American Teche

An international conflict of enormous consequence erupted over a thousand miles from the languorous calm of Bayou Teche: the American Revolution, which pitted the newborn United States of America and its ally, France, against the British Empire. In 1779—the same year that Bouligny founded Nueva Iberia and hired Gonsoulin and Grevemberg to explore the bayou—the French persuaded Spain to join the conflict and attack British strongholds on the Gulf Coast. In New Orleans, Gálvez responded with celerity, raising a small but committed army eager for combat with an age-old enemy.

In the Teche Country many responded as required to the governor's call to arms. From the banks of the bayou came a disparate force of French and Spanish settlers, Acadian exiles itching to avenge the expulsion, *gens de couleur libre*, and even armed black slaves. Bouligny himself left Nueva Iberia with a force of five professional soldiers and a mix of retired veterans, militiaman, repentant deserters, Malagueño settlers, far-flung Americans, twenty-five slaves, and others. The local commandant, DeClouet, led additional Teche Country settlers into the conflict. His militia officers and enlisted men bore surnames well-known along the bayou: Grevemberg, Delahoussaye, Broussard, Dugas, Bernard, Boutté, LeBlanc, Prevost, Prejean, Guilbeaux, Doucet, Thibodeaux, Trahan, Hebert, Boudreaux, Wilse, Robichaux. These soldiers participated in the American Revolution as much as any New England minuteman and helped eject the British from the Gulf Coast—and did so again in 1815, when a similar multiethnic amalgam (including Captain Joseph Dubuclet's volunteer

Hussars of the Teche) defeated an invading British force at the Battle of New Orleans.[1]

In the meantime, Napoleon Bonaparte coaxed Louisiana from the Spanish crown in 1800 and three years later sold the vast territory to the United States. In 1804 an American officer, Lieutenant Henry Hopkins, arrived in the Attakapas region to formally raise the U.S. flag over the Teche. The bayou was now American—legally and politically, if not culturally.[2]

By then a diverse blend of racial and ethnic groups populated the banks of Bayou Teche: French, Spanish, Acadian, African, even German and Irish, along with the indigenous Chitimacha. Among them, however, lived an ever-growing number of Anglos, particularly on the lower Teche in St. Mary Parish. These Anglos brought to the bayou the English language, various strains of Protestantism, and a more commercial worldview. "Money, Negroes, sugar, and cotton and land seems to engross all their time and attention . . . ," a visitor to Anglo-populated Franklin noted around 1820. This outlook would shortly transform the Teche from a sleepy frontier stream into a thriving economic thoroughfare—a veritable Mississippi River in miniature.[3]

Pioneers of Anglo stock had appeared on the Teche as early as 1773. That year Gabriel Fuselier de la Claire, serving as commandant, recorded the presence of three Anglo settlers in the Attakapas and Opelousas districts. They included two unnamed carpenters and a certain John Hamilton, a merchant conducting trade between the Teche Country and New Orleans. Six years later Bouligny noted two Anglo residents at Nueva Iberia, namely, the aforementioned Thomas Berwick and William Henderson, who helped the Malagueños build their colonial outpost on the bayou at Petite Fausse Pointe.[4]

Not until 1780, however, did Anglo settlers appear in the Teche Country in sizable numbers. Several trends and events, observed historian Glenn R. Conrad, spurred this migration. As Conrad explained, "Some were running from the American Revolution, and thus might be considered political refugees; some were fleeing religious intolerance; others were escaping from generations of ethnic degradation; but all sought the chance for a new start in life which only the frontier could provide."[5]

These newly arrived Anglo frontiersmen hailed from Massachusetts, Pennsylvania, Maryland, Virginia, and the Mississippi territory, among other English-speaking regions. Unable at first to afford their own land, they tended cattle for well-to-do French Creole landowners. Others worked as cobblers, saddlers, and tanners, whom the Creole planters engaged not only to produce goods, but to train slaves in these crafts. Some Anglo settlers put their educations to work, serving as tutors to planters' children. Still others used carpentry skills to build dwellings, cotton gins, and sugar mills, or drew on commercial skills to become, as Conrad noted, "for all practical purposes, the area's first general merchandisers." A few of these merchants generated enough capital to buy schooners, which carried the Teche Country's produce to New Orleans and returned with consumer goods from "the City."[6]

Anglo pioneers eventually bought their own farmland and joined the emergent planter class along the bayou. They did so, however, primarily on the lower Teche, creating an Anglo enclave that stretched southeast from a nebulous point below present-day Jeanerette to the bayou's mouth, a distance of about forty-five miles by water. By 1840 Anglos made up about 60 percent of the white population along the lowest stretch of Bayou Teche. In contrast, they made up only about 12 percent of the white population upstream in neighboring St. Martin Parish, where French-speaking Acadian, French, and French Creole landowners dominated the cultural landscape. Yet as one historian aptly noted, the Anglos, whether on the upper or lower Teche, "were far more important than their numbers indicate," because they more than any other group introduced the Teche Country to modern agriculture, commerce, and transportation.[7]

As the region embraced the American concept of progress, new communities sprang up along the Teche. By the late antebellum period they numbered over a dozen. From the bayou's headwaters to its mouth, settlements large enough to warrant a U.S. Post Office by the 1850s included Barre's Landing (now Port Barre); Leonville; Arnaudville; Breaux's Bridge (Breaux Bridge); St. Martinsville (St. Martinville); Fausse Pointe (Loreauville); New Iberia; Jeanerette; Charenton; Franklin; Centreville (Centerville); and Pattersonville (Patterson). Of these

A postcard image of the Teche at St. Martinville, ca. 1920. The chimneyed brick structure near center is the Old Castillo hotel. To the right of it is St. Martin de Tours Catholic Church and the Evangeline Oak. Water hyacinths in bloom line the banks. Source: author's collection.

communities, the most populous, affluent, and reliant on the Teche for prosperity were St. Martinsville, New Iberia, and Franklin.[8]

Myth clouds the origin and early development of St. Martinsville (a spelling that persisted into the early twentieth century). Contrary to popular belief, the town was not founded in the mid-1700s as a trading post called *Poste des Attakapas*. (In actuality, that term applied to the entire roughly 3,000-square-mile district.) Moreover, it seems there was nothing at the present-day site of St. Martinville, except a few slaves and *engagés* (indentured servants) tending cattle, until 1773. That year colonists hired settler Jean Berard to oversee construction of a church on land donated by Dauterive, the retired French military officer and absentee landowner. On Dauterive's death his widow sold most of the remaining land to a handful of settlers. Still, by the late eighteenth century little stood on these tracts except a few farm dwellings and outbuildings and the completed church. Then, around 1800, two settlers sold off land in lots to establish a town. By 1806 Governor Claiborne could refer to visiting "the county town of Attakapas." What comprised the town, he did not record, but eleven years later a correspondent wrote of the town, by then formally called "St. Martinsville":

> It contains about 40 dwelling houses, besides out-houses. It has one Roman Catholic church and residence for a priest, one court house, and a public jail, one academy, a small market house for meat only, three taverns, three blacksmiths' shops, two hatters' shops, three tailors' shops, one saddler's shop, two boot and shoemakers' shops, one joiner's shop, one silversmith's shop, two bakers' shops, one tinner's shop and ten stores. . . . [along with] four attorneys at law and three physicians.[9]

Approximately twenty miles south by water lay New Iberia, formerly Nueva Iberia. After its 1779 founding the village had fared poorly and, as Conrad bluntly asserted, "was nothing short of a complete failure." The problems were manifold: the Malagueños sowed crops unsuited to the climate; basic supplies chronically ran short; and the settlers grew complacent on resulting government subsidies. In the mid-1780s some of the Malagueños abandoned the bayouside settlement; the others melted away after 1795, some of them moving on to the prairie near Lac Flamand (which accounts for its current name, *Spanish Lake*).[10]

Within a few years, however, French Creoles and Anglos had bought up the site of Nueva Iberia, along with adjacent property on the Teche. The village became *New Iberia*, less commonly *Nova Iberia* and *New Town*—the latter possibly meant to distinguish the community from the extinct "old town" founded by the Spanish. An 1817 visitor observed of the revived village, "New Iberia is quite a small place, only one private house and a tavern, and the tavern serves for post office, custom house, and tavern." By the 1850s, however, as many as 250 settlers lived in the now-thriving town.[11]

About thirty-seven miles southeast of New Iberia by water sat Franklin, an Anglo community in francophone south Louisiana. One of these Anglos, Alexander Lewis of Tennessee, founded Franklin. In 1814 Lewis bought up about 436 arpents (roughly 369 acres) on the east and west banks of the Teche. He donated plots for a courthouse, jail, school, and other public entities, and offered his remaining land as lots for sale. As an observer noted around 1815, "Franklin contains

A steamboat on the Teche unloads at New Iberia, 1874. Source: *Frank Leslie's Illustrirte Zeitung* (1874), Prints, Sketches, and Poster Collection, Coll. 182, University Archives and Acadiana Manuscripts Collections, Edith Garland Dupré Library, University of Louisiana at Lafayette.

about twelve or thirteen dwelling houses, a tavern, and a jail. The houses are generally indifferent. The tavern when finished will be a tolerable good house. It is kept by a Mr. Reed, a native of Virginia. . . . They have one house here of but mean appearance which serves alternately as court house, church, and school house." (Reed's tavern, also known as Hulick's tavern, still stands behind the antebellum home called Shadowlawn.)[12]

St. Martinsville, New Iberia, and Franklin, as well as the other, smaller communities that lined the Teche, shared at least one obvious trait: they came into being because of the region's plentiful natural resources—namely the Teche itself and the rich, dark alluvium that formed its banks. After cattle ranching, landowners along the Teche tried their hand at cash crops. One of the earliest, as mentioned, was indigo. Flax, hemp, corn, and tobacco were also tried in a small way, but, smothered by the region's semitropical climate, failed to become major crops. Landowners at last found success in cotton and sugarcane. As geographer Lauren C. Post accurately noted, "There had been some competition between cane and cotton for land, and the lower Teche went primarily to sugar cane. The upper Teche . . . went in for cotton."[13]

Settlers, however, originally planted cotton on both the upper and lower Teche. It became an important crop by 1769, when Kelly and Nugent observed, "From cotton [the Acadians] make very good cloth for their clothes." The commodity still dominated the Teche Country economy in the early nineteenth century. As historian Carl A. Brasseaux noted of the Acadians during that period:

> Raw cotton was either cleaned by hand or fed through pre-Whitney cotton gins, which were quite common. . . . The processed fiber was then spun into thread; the spools of thread were mounted on the ubiquitous household looms and woven into *cotonnade*. This rough fabric, which was usually left in its natural [yellow-brown] state or dyed indigo blue (though other colors were sometimes used), was then fashioned into clothing for the entire family: pants, *garde-soleils* (sun bonnets), shirts, blouses, *carmagnolles* (short, decorated vests), dresses, and floor-length skirts . . . [as well as] cotton stockings. . . .[14]

An early-nineteenth-century survey of the upper Teche Country identified only two industrial (non-farming) trades: hide-tanning and the production of cotonnade. Cotton likewise dominated the lower Teche, whose planters had not yet embraced sugarcane. Judge Wilkinson thus recalled, "The only staple cultivation in 1810, for market, was cotton."[15]

By that decade, however, some lower Teche planters were already experimenting with sugarcane. The spindly stalks seemed a viable alternative to cotton, which in the bayou's lower reaches frequently suffered from "worms and the rot"—armyworms and an equally devastating microbial disease. Although in coming generations lower Teche farmers would try their hand again at cotton, and even at large-scale rice production, sugarcane would become their unrivaled chief commodity.[16]

Sugarcane came to the lower Teche as early as 1766, when Acadian exile Jean-Baptiste Semer implied that his fellow Acadians were producing sugar along the bayou. The exiles probably grew a trifling for their own consumption, and decades would pass before New Orleans entrepreneur Etienne de Boré demonstrated the crop's large-scale profitability in south Louisiana. By the early nineteenth century a sugar mill had appeared on the banks of the Teche. As Judge Wilkinson recollected of that era, "There was one sugar plantation in the county, and that was at St. Marc Darby's [Jean-Baptiste St. Marc Darby], and the cane was ground by water power." Sitting on the west bank of Bayou Teche above New Iberia, Darby's plantation (or more properly that of his widow) did indeed boast a watermill for grinding sugarcane—an unusual source of power in south Louisiana's flat coastal region. Grasping that nearby Spanish Lake had a higher water level than the Teche—a sixteen-foot difference, according to one source—the Darbys ran a canal between the lake and bayou, and used the downward flow of water, probably in conjunction with a dam, to turn a waterwheel. (Another waterwheel, apparently built by St. Denis DeBlanc, stood a short distance up the Teche at Keystone Plantation and also used water from Spanish Lake for propulsion.) "The idea of making sugar in those days, in Attakapas," continued Wilkinson, "was looked upon as visionary" as well as "with distrust." Others gradually planted the peculiar crop, however, and, as the Judge recorded, "It led . . . to a very important result: it proved successful, and several persons of limited means were induced to embark in the business. . . ."[17]

Regardless, it was not until around 1825 that sugarcane became a common crop on the lower Teche. In 1827 an observer noted, "Above St. Martinsville, cotton is universally cultivated on the banks of the Teche: below it, are a number of sugar plantations, which succeed remarkably well." A half-century later sugar planter F. D. Richardson of Bayside Plantation (whose big house still overlooks the Teche at Jeanerette) related: "At that date [1829] we recall but few sugar houses on the public road [that followed the bayou] from Franklin to New Iberia, to wit, Agricole Fuselier, Dr. Solonge Sorrel, Fréderic Pellerin and Nicholas Loisel; but in 1835 nearly all the plantations on the Teche were in sugar. Those six years had done the work of a generation in changing the staple commodity of a country, its implements of husbandry, and in many respects the habits and customs of a whole community."[18]

By the mid-1840s the lower Teche supported a total of 103 sugarcane planters. Their ranks included 11 Acadians, 50 Frenchmen or French Creoles, and 36 Anglos. These planters exhibited a clear settlement pattern: the farther one descended the bayou toward its mouth, it became less Acadian and French Creole and more Anglo. For example, of the Acadian planters all but one resided upstream in St. Martin Parish, while the Creole planters resided primarily in St. Martin and St. Mary parishes above Franklin (though some also lived below Franklin). Of the Anglo planters, however, all but three resided in St. Mary Parish, settling evenly above and below Franklin. (Iberia Parish is not mentioned because it was created only later, in 1868, from parts of St. Martin and St. Mary parishes.)[19]

Interestingly, at least two Creole of Color planters lived on the lower Teche in St. Mary Parish. Occupying a middle ground between free whites and enslaved blacks, Creoles of Color (*gens de couleur libre* or "free persons of color") descended from mixed-race ancestry. They hailed from a blend of black and white, sometimes black, white, and Native American lines. Among other privileges, Creoles of Color could own property, including slaves. Thus, one of these Creole of Color planters along the Teche, Jean-Louis Senette, estimated his worth in 1850 at a respectable $9,000, a sum that included the value of his nine black slaves and one who was, like Senette himself, a mulatto. The other Creole of Color planter along the Teche, Romaine Verdun, estimated his worth that year at $15,000 and counted twenty-four slaves, five of whom were mulattos like himself.[20]

With the rise of sugar plantations along the lower Teche came sugar houses, also called sugar mills. Often made of local red-orange brick, they were square or rectangular structures, or complexes of a few adjacent structures, from which sprouted one or two smokestacks. Each plantation operated a sugar house, which became a hive of activity during the harvest, known as "the grinding season" or, in local French, *la roulaison*. This French term derived from the verb *rouler*, meaning "to roll," because slaves fed the hand-cut cane through rollers that squeezed out the raw saccharine juice. In early days animals turned these rollers, but by the Civil War nearly eight in ten sugar houses used steam-driven rollers. Likewise, primitive iron kettles that boiled cane juice into syrup eventually gave way to state-of-the-art evaporators, filters, vacuum pans, and centrifuges.[21]

Despite the advent of this new technology, sugar houses continued to depend on slave labor. Operating incessantly from October through January, the mills used slaves to feed cane into the rollers, chop firewood for boilers, and perform myriad other tasks involved in making sugar. F. D. Richardson romanticized the slaves' grueling work and ungodly hours, recalling, "[T]here was something inspiring about a caneyard at night, all illuminated, that kept [the slave] in a merry mood. . . . His jokes and loud, ringing laugh kept time with the rattle of the cane as he dashed it on the carrier and wheeled to get another turn." Richardson added, "[T]here were pleasures for the Negroes about a sugarhouse unknown to cotton plantations. There is sugar cane, to begin with, and no shifty darky would be without a stalk to his mouth pretty much all the time he had to spare, and most of them did not wait for spare time. Then there was hot juice to be drunk, with now and then a chance at the strike box, and trough candy, with taffy and molasses *ad libitum*."[22]

At the end of the sugar-making process slaves grappled with the unwieldy, oversized barrels called hogsheads, used for sending granulated brown sugar and molasses to market. (They produced none of the granulated white sugar found today in supermarkets.) Each hogshead weighed approximately 1,100 pounds, and in 1844, for instance, sugar planters on the Teche produced 13,295 hogsheads of raw sugar—in other words, over 15 million pounds of the commodity. And this count excludes sugar made by planters living near but not directly on the Teche.

Regardless, there was only one way to get all that sugar and molasses to market: by boat, on the bayou. Although some boat captains descended the Teche and Lower Atchafalaya to navigate the open waters of the Gulf of Mexico, others preferred inland routes. These passages ran from the Teche through the massive Atchafalaya swamp to enter the Mississippi River via, for instance, Bayou Plaquemine or the Attakapas Canal (actually an enlarged natural waterway).[23]

Just as early Teche Country cotton planters did with their fleecy bales, early sugarcane planters sent their hogsheads to market using keelboats. These were long, narrow vessels pushed with wooden poles. (Their single mast amidships could not be used on the bayou.) "[A] number of what they call keel boats pass Franklin everyday down the Teche," surveyor John Landreth recorded from that village in 1819, "carrying from one hundred to three hundred bales of cotton each. These boats are generally rowed by eight, ten, and twelve oars and [also have] a man to steer."

Deep-water sailing vessels also ascended the bayou, delivering goods from as far as the upper East Coast and taking away sugar, molasses, and other local products. St. Martinsville lay too far up the Teche for these sizable ships, but New Iberia sat on deep enough water to accommodate "vessels of considerable burthen," as surveyor William Darby remarked in 1817. In 1819 the Collector of Customs for New Iberia wrote of "[t]he schooner *James Lawrence* . . . now at New Iberia"—a vessel that when fully loaded drew six feet of water, accommodated up to ten passengers (excluding the crew), and shipped about thirty-three tons of cargo. Likewise, a journalist visiting Franklin in 1853 witnessed on the Teche: "Vessels of large size . . . [that] mingle their rigging with the foliage of the forest. Here these argosies [large merchant ships], born in the cold regions of the Aroostook [a river in Maine and Canada], fill their holds with sugar and molasses, and, once freighted, wing their way to the north." These tall-masted ships would have ascended the bayou not by using their sails, but by poling, warping (pulling a vessel by hand or winch using a rope tied to a tree or other immobile object), cordelling (towing a vessel by rope using human or animal power), and, most primitively, bushwhacking (propelling a vessel manually by pulling on overhanging limbs and branches). As late as 1890 the writer Lylie O. Harris noted on a Teche outing, "The schooners and red-and-white sailed luggers . . . cordelled up stream as though man were yet a beast of burden. . . ."[24]

So many of these seafaring ships called on the Teche Country that in 1811 the federal government created a customs zone, named the District of the Teche, and selected New Iberia as its official port of entry. Agents there kept track not only of vessels coming and going, but fought to prevent smuggling on the bayou and the various bays and inlets along the nearby coast. A visitor to New Iberia in 1819 thus observed on the Teche "several small vessels laying here, some of which . . . [have been] seized in the disgraceful smuggling trade carried on in this country." That smuggling trade included trafficking in foreign silk, alcohol, and tobacco to avoid tariffs, and human trafficking in African slaves, whose import Congress had outlawed in 1808. A New Iberia customs agent, for instance, reported in 1817 that his deputy had seized "twenty-three African Negroes, which had been introduced into my district . . . by way of M'Call's Island [Petite Anse Island], Petite Auce [Anse] Bayou. . . ." (Tales have circulated for generations about pirates using Bayou Teche as an entrepôt for their illegal smuggling activities. As a journalist noted in 1855, "Some of the old Creoles along Bayou Teche remember to have witnessed the traffic which the pirates carried on with the settlers, exchanging rum and dry goods for the provisions of the country. . . ." A few Teche Country residents—including François Mongault, whose house stood in New Iberia on the bayou—were supposedly loyal agents of famed Louisiana pirate Jean Lafitte. Such claims, however, remain unconfirmed.)[25]

New Iberia, however, had a rival in Franklin, which in 1830 became the district's new port of entry. By 1846 the government employed both a Collector of Customs and an Inspector of Revenue at Franklin, which traded via the Teche with New York, Philadelphia, Baltimore, Portsmouth, Charleston, and New Haven, among other major seaports. Over a nine-month period in 1842–43, ninety-one seafaring vessels left Franklin with over 5,350 hogsheads or barrels of sugar, over 9,200 hogsheads or barrels of molasses, and over 81,700 feet of live oak (as well as, less impressively, 95 bales of Spanish moss, used to stuff mattresses and pillows).[26]

By then keelboats and sailing ships had already been joined by a new type of vessel, one that revolutionized transportation on the Teche: the steamboat.

Steamboats first appeared in North America in 1790, when John Fitch operated a primitive steam-powered vessel on the Delaware River. Seventeen years later Robert Fulton and business partner Robert R. Livingston offered commercial steamboat service on the Hudson. In 1812 steamboats first navigated the Mississippi; seven years later over thirty such vessels plied that river and its tributaries. Unlike earlier, smaller forms of water transportation relying on muscle or wind for propulsion, steam-powered vessels provided a swift, reliable means of moving large quantities of cargo and passengers, even against strong

The steamboat *Black Prince* on the Teche near St. Martinville, ca. 1900. The bayou at this place was not much wider than the vessel. Firewood sits on deck, presumably for the ship's boilers. A person stands on the wooden ferry at right. Source: author's collection.

currents. Moreover, steamboats could maneuver deftly even on narrow, winding waterways.[27]

These attributes made the steamboat ideal for south Louisiana's snake-like bayous, and Teche Country entrepreneurs took notice. In 1819 several planters on the bayou received state backing for "the establishment of a line of steamboats, to ply between the Rivers Vermilion and Teche and the Mississippi." Incorporated that year, the Attakapas Steam Boat Company built a 295-ton vessel aptly christened the *Teche*, which ran from the bayou to New Orleans via the open waters of the Gulf of Mexico. Within a few years, however, the company failed, and the *Teche*, sold to another firm, exploded on the Mississippi River in 1825, killing twenty.[28]

That same year a relatively small vessel, the 48-ton *Louisville*, proved that a steamboat could reach the Teche from New Orleans via the same inland route used by smaller, more primitive boats. Veering from the Mississippi River into Bayou Plaquemine (no minor feat given the treacherous current there), the *Louisville* crossed the Atchafalaya

The Steamboat *John D. Grace* at New Iberia, ca. 1910. Like many steamboats on the Teche, it would not have been able to turn around unless it arrived at a turning basin carved out of the banks. Source: author's collection.

swamp, entered the Atchafalaya and Lower Atchafalaya rivers, and steered up the Teche. This accomplishment signified a reduction in travel time and shipping costs, factors of great importance to Teche Country planters. St. Martinsville fêted the vessel's crew, and on April 16, 1825, the town's *Attakapas Gazette* newspaper reported: "The steamboat *Louisville*, [under] Capt. Curry, very unexpectedly made her appearance in the river Teche on last Monday, having entered by way of the Bayou Plaquemine. This arrival forms a new era in the history of our section of country. Hitherto the prevailing opinion has been that the navigation between Plaquemine and the mouth of the Teche was impracticable for vessels of this description."[29]

Other steamboats soon appeared on the Teche, including both packets (which came and went according to regular schedules) and transients (or "tramps," as they were also called, which had no fixed schedules). By 1829, for example, six steamboats plied the meandering route between the bayou's inland ports and New Orleans. Two decades later, in 1850, steamboats from the Teche Country docked at New Orleans no less than 153 times (some boats no doubt making the same run repeatedly).

By 1857 the Teche district boasted 2,225 tons of steam vessels, but how many individual vessels accounted for this figure is unknown.[30]

A partial steamboat listing indicates that well over a hundred vessels operated on the Teche in the decades prior to the Civil War. Although the number of boats plying the Teche at any given time is undetermined, eyewitness accounts suffice to confirm that "numerous steamers pass up and down its gentle stream" (to quote a late-antebellum British visitor). These vessels ranged from small steamers like the *Houma*, which weighed almost 56 tons and measured 97 feet in length and 26 feet in width, to behemoths like the *Tomochichi*, which weighed 236 tons and ran 150 feet in length and 28 feet in width. (The Teche itself is only about 130 feet wide at St. Martinville—meaning the largest vessels could turn around only by pulling into artificial cuts, called "turning basins," made in the banks for this purpose. Some of these basins are still visible today along the bayou.)[31]

These steamboats often bore colorful names, such as the *Belle Créole*, the *Arabian*, the *Louis d'Or*, the *Swiftsure*, and the *Velocipede*. Many names honored persons now largely forgotten, such as the *R. C. Oglesby*, the *T. D. Hine*, the *Major Aubry*, the *General Morgan*, and the *D. B. Mosby*; or, in a feminine vein, the *Mary Bess*, the *Mary Jane*, the *Naomi*, the *Rosa*, the *Anna*, and the *Jenny Lind*. Although some of these boats were constructed locally, many hailed from shipyards in seemingly far-flung places like Ohio, Indiana, and Pennsylvania, a detail testifying to the declining insularity of the Teche Country.

Steamboats broke down barriers by connecting the bayou more directly to the outside world. From New Orleans they brought flour, pork, whiskey, dried apples, mackerel, furniture, jewelry, hardware, and clothing; and from the Teche Country they carried away sugar, molasses, rum, cattle, cotton, lumber, potatoes, oranges, peaches, eggs, pelts, and Spanish moss. (Interestingly, before the railroad's arrival in 1879 Teche steamboats would carry to market a fiery new product called Tabasco brand pepper sauce. Introduced in 1868 by former banker E. McIlhenny of Avery Island, Tabasco sauce relied on the bayou's steamboat trade to reach the major shipping hub at New Orleans and, through it, other sections of the United States and parts of Europe.) Steamboats also carried mail, news, and people up and down the Teche, into and out of the region. As Judge John Moore of St. Martin Parish observed in December 1851, "I have ascertained that the steamboats navigating the Bayou Teche alone have carried, since the 8th of January last, about six thousand passengers, *to* and *fro*."[32]

As routine as steamboat travel became on the Teche, it could be dangerous, as on other rivers. High-pressure boilers exploded, often because captains pushed them beyond capacity to gain extra speed or power. A sunken log or other "snag" could easily damage, even sink, a steamboat, especially in turbulent waters—for while bayous usually flow sluggishly, they can run swiftly after a deluge. As historians Carl A. Brasseaux and Keith P. Fontenot have observed, the bayous were "Filled with floating debris, bristling with submerged navigational hazards, and fraught with treacherous currents." Of the Teche itself, they noted:

> By 1850, [the bayou] was riddled with shoals caused by siltation from drainage canals on the many sugar plantations lining the stream; in many places, the stream was dangerously shallow. This shoaling problem was so severe that in 1857 the state legislature appropriated $11,250 for construction of a lock on Bayou Teche [built only in 1913]. In the late 1850s, a variety of hazards—sunken logs, floating 'mill logs,' 'immense quantities' of floating grass which accumulated into floating islands, the sunken remains of discarded flatboats, and public and private bridges—collectively made late antebellum steam navigation on Bayou Teche increasingly perilous.

To counter these dangers, the state of Louisiana sent "snag boats" to clean the maze of congested bayous. Manned by large crews, from captains, pilots, and engineers down to cooks, carpenters, and blacksmiths, these vessels patrolled the waterways armed with chains, ropes, steam-driven winches, and gangs of slaves expert with the axe and crosscut saw. Superintendent of the snag boat *General Walker*, Jean Jules Hardy of Breaux Bridge provided a glimpse into their operation when he recorded in April 1853:

> After having completed the work of B[ayou] Cyrus, I proceeded to Bayou Teche, where I am now working since this morning. I have commenced to work twelve miles above Breau's Bridge [*sic*], at Lastrapes' plantation going up. The work of

> B. Teche is very heavy on account of half of the leaning trees are live oak. There is about twenty-six miles to clean out in B. Teche, to complete that bayou. Six miles in Parish St. Martin, and twenty in St. Landry.
>
> The planters are highly satisfied to see the state hands working in the bayou, where the state never had cut a tree down. . . . I am now cutting the leaning trees on the west side of B. Teche.[33]

Clearing the bayou was a thankless ordeal for Hardy's state-owned slaves. When not cutting down low-leaning trees, they hoisted sunken logs, pulled up stumps, or cut firewood to feed the snag boat's boilers. They also contended with leaks, rotting ropes, low provisions, illnesses, injuries, summer heat, snakes, alligators, mosquitoes, and the semi-tropical region's frequent rainstorms. "The water, it is too high to work in the Teche," complained Hardy in a typical entry. "Weather, rain, rain, every day. Bad weather."[34]

Hardy and his enslaved workers could not have known that Louisiana's rivers and bayous, including the Teche, served as conduits for spreading disease-carrying mosquitoes. That disease, yellow fever, manifested itself in high fever, chills, nausea, and vomiting, yellowing of the skin (hence the name), and hemorrhaging from the eyes, ears, nose, mouth, and other orifices. In critical cases, victims suffered delirium and convulsions before slipping into a coma, and finally dying.[35]

Although malaria, cholera, and typhoid were known to infect Louisiana residents, a more feared disease, yellow fever, erupted periodically. It often arrived in New Orleans from other Caribbean-rim ports before spreading to the interior. Although historians regard a New Orleans outbreak of 1796 as Louisiana's first confirmed yellow fever epidemic, the disease may have ravaged the Teche Country as early as 1765. That summer and fall an unidentified illness, described by a survivor as a "fever," devastated the newly settled Acadian exiles along the bayou. By the epidemic's end, approximately one in five of the nearly two hundred exiles had perished. Still, some other disease, such as smallpox or typhoid, might have accounted for those deaths.[36]

In any event, yellow fever had definitely reached the Teche Country by the early nineteenth century, when the advent of the steamboat era in the 1820s and 1830s magnified the disease's impact. Steamers spread the sickness deeper into the interior, with a swiftness previously unknown. Moreover, south Louisiana's sugarcane region encouraged both mosquito reproduction and disease transmission. The crop thrived in a hot, humid environment and its cultivation relied on bayous for transportation as well as on ditches and canals (such as those forming a watery labyrinth in coastal lower Iberia and lower St. Mary parishes) for crucial drainage. These factors afforded mosquitos an ideal habitat for proliferation. Plantations also brought human beings together in compact populations able to support outbreaks, for the disease could not flourish without a critical mass of potential hosts.[37]

As a result, the late antebellum period witnessed a significant upsurge in yellow fever outbreaks on Bayou Teche. An epidemic, for instance, struck Franklin, St. Martinsville, and New Iberia in 1839. It was during this outbreak that a black Santo Dominican woman named Félecité—possibly immune to the disease through prior exposure—cared tirelessly for infected New Iberia residents, regardless of color. As a historian recorded in 1891, "She nursed the sick, administered to the dying, closed the eyes of the dead, and wept over their graves." (Because of her bravery and kindness, Félecité remains a venerated figure in modern-day New Iberia.)[38]

The 1850s, however, witnessed a series of catastrophic outbreaks along the Teche: in 1853 the pestilence hit Franklin, Centreville, and Pattersonville; in 1854, Franklin and Pattersonville; in 1855, St. Martinsville, New Iberia, Centreville, and Pattersonville. The first of these epidemics came to the Teche on August 8, 1853, when a family of five northern emigrants—a father, mother, son, and two daughters—debarked from the steamboat *P. Miller* a little upstream from Pattersonville. "On the second day," recorded the attending physician, "the eldest daughter, a young lady of nineteen years, was taken with black vomit; but lingered forty-eight hours and died. The young man was taken with black vomit some twelve hours after his sister, and died in ten or twelve hours after that symptom supervened. The younger sister recovered, after a serious and protracted illness."[39]

While she recovered, the illness spread throughout the countryside and to Pattersonville itself. "In our little village," continued the doctor, "nearly every individual had the disease, during some period of its prevalence—that is, from the above date [September 21] until late in December. Whenever it made its appearance in a family, it generally sooner or later extended to every member, in what we considered the infected region." Indeed, the population of Pattersonville totaled 600 persons, yet the number of infected in the town's locale—regarded as a roughly fifteen-mile stretch of the Teche—numbered over 500, of whom 45 died.[40]

The illness spread upstream to Centreville on September 15, when it struck "a mulatto man, a cooper, working and sleeping in a shop on the banks of the Teche," as a town physician recorded. "He died on the ninth day, after a relapse on the fifth, having hiccough[s], haemorrhage from the gums and nose, spasms, and yellowness of skin and eyes." The doctor could identify no source for the fever, though he eyed with suspicion the cooper's refuse, "the collection of decaying [wood] chips and shavings on the bayou. . . ." In the following weeks the disease swept through Centreville. Physicians ultimately diagnosed 45 cases among the town's roughly 200 inhabitants, and of the infected 7 died (4 whites and 3 blacks).[41]

At Franklin, where the disease arrived on October 19, only 5 of the town's 1,400 residents became infected—a figure one local doctor attributed to "our famous quarantine." That quarantine consisted of a guarded checkpoint on the bayou road a half-mile below town; and two quarantine stations staffed by physicians, also situated below town. These stations examined all approaching riverboats and "the mail boat and vessels coming in from sea." Boats hailing from infected regions underwent a nine-day quarantine, dating from their day of embarkation.

Finally, for good measure Franklin set up two cannons below town and trained their muzzles on the bayou.

This *cordon sanitaire* did not sit well with the captain of the *P. Miller*—the same vessel that earlier introduced the disease to Pattersonville—who despite the quarantine steered his boat toward Franklin. His steamer dodged a warning shot from the cannons, but "seeing a pretty formidable posse collected on the wharf . . . ," as Franklin's doctor recalled, "she returned to quarantine ground, discharged the portion of her cargo intended for this and other places on the Teche, and recrossed the Lake [Grand Lake]." Of the town's refusal to receive him, observed a local paper, "The captain swore lustily . . . but to no purpose. . . ."[42]

Like the 1853 pestilence, the yellow fever outbreaks of 1854 and 1855 wafted through the swamps to grip the Teche Country. Again the mosquito-borne virus struck many, and to some brought a horrible death. "The health of our town, we are sorry to say, has not been very good . . . ," lamented a St. Martinsville newspaper in fall 1855. "We are placed under the painful necessity of announcing the demise of several of our citizens . . . the disease being of a malignant putrid fever which terminates in black vomit." It was yellow fever, as the paper no doubt well knew, even while its report avoided mention of the dreaded sickness by name.[43]

A more devastating catastrophe, however, came shortly to the Teche. It brought violence, looting, poverty, and hunger. It destroyed elegant plantation homes, wiped out crops and livestock, filled many forgotten graves, and clogged the bayou with collapsed bridges and rotting, half-sunken vessels. That catastrophe was the tragic four-year conflict known as the American Civil War. Yet this struggle would lead to freedom for millions of enslaved African Americans, including thousands of field workers and house servants who toiled in captivity up and down the Teche.

4

The Teche during Wartime

From 1861 to 1865 the American Civil War divided the United States, pitting the anti-slavery North against the pro-slavery South. This bloody struggle killed hundreds of thousands of Americans, who fought each other in battles like Bull Run, Shiloh, Antietam, Chancellorsville, and Gettysburg. Although far from these major Civil War battlefields, the Teche country experienced its share of wartime turmoil. After Union troops captured New Orleans, Baton Rouge, the lower Mississippi River, and the Bayou Lafourche region, they took Brashear City (present-day Morgan City), crossed Berwick Bay, and swept up Bayou Teche. Both Union and Rebel forces regarded the Teche as vital to their respective war efforts. For the South, the bayou held enormous wealth in sugar, cotton, and slaves; for the North, it offered an invasion route directly into the state's fertile interior.[1]

In fact, the Teche country endured three Union campaigns and two occupations, each beginning at Brashear City, where Union forces massed before marching up the waterway. The first Teche invasion occurred in spring 1863, when Union troops ascended the bayou to cut supply lines between Texas farmlands and under-provisioned Rebels on the Mississippi. Later that same year Union forces launched the second Teche invasion, going up the bayou with the goal of pivoting west to assault Texas. Dogged by resistance, however, they got no farther than Carencro before falling back to New Iberia, Franklin, and eventually Brashear City. The third Teche invasion occurred in spring 1864. This time Union troops returned as a prelude to the Red River Campaign—an offensive that ended in their humiliating defeat in north Louisiana.

All three campaigns found General Nathanial Banks commanding the Union forces and General Richard Taylor, son of U.S. president Zachary Taylor and former Stonewall Jackson lieutenant, leading the Rebels.[2]

Although secondary to the more renowned events of the Civil War, these invasions tied the bayou's fate to that of the not-too-distant Mississippi—"the all-important object of the present campaign," wrote Henry W. Halleck, general-in-chief of all Union armies, "worth to us forty Richmonds." As a Union officer who fought along the bayou remarked:

> The Teche country was to the war in Louisiana what the Shenandoah Valley was to the war in Virginia. It was a sort of back alley, parallel to the main street, in which the heavy fighting must go on; and one side or the other was always running up or down the Teche with the other side in full chase after it. . . . But why should [we] go by the back alley of the Teche instead of by the main street of the Mississippi? Because it was necessary to destroy the army of [Taylor], or, at least, to drive it as far as possible, in order to incapacitate it from attacking New Orleans while we should be engaged with the fortress of the bluffs [Port Hudson, which with Vicksburg denied total control of the Mississippi to the North].

Even after the Union secured the entire Mississippi in summer 1863, Taylor's army remained a significant threat to Banks's operations. This partly explains why in less than two years' time the Union general launched three campaigns up the Teche.[3]

Union forces first entered the Teche Country in early November 1862, when four U.S. gunboats—the *Calhoun*, *Estrella*, *Diana*, and *Kinsman*—steamed up the bayou in search of the Rebel gunboat *Cotton*, a side-wheeled partial ironclad armed with four guns. Days earlier the *Cotton* had audaciously clashed with Union vessels on Berwick Bay, only to escape by fleeing up the Lower Atchafalaya into the Teche. "As soon as I had coaled I started with all four boats up the [Lower] Atchafalaya River," wrote Lieutenant Commander Thomas McKean Buchanan, commander of the *Calhoun*, "to go up Bayou Teche to Franklin." Well short of Franklin, however, the Union flotilla found the bayou obstructed by felled trees and scuttled vessels at a spot called Cornay's Bridge. Above these obstructions waited the *Cotton*.[4]

The CSS *Cotton* battles the Union gunboats *Kinsman, Estrella, Calhoun,* and *Diana*. Only three of the Union vessels are shown; the *Cotton* is in the background, flying a Confederate naval flag. The troops onshore are Rebel infantry, artillery, and cavalry. Source: *Harper's Weekly* VII (February 14, 1863).

The *Calhoun* opened fire, but its gun carriage immediately broke loose. Buchanan's crew dashed to make repairs while the other Union vessels moved to the fore. Meanwhile, the *Cotton* returned fire and disabled the *Estrella*, which "was obliged to run on shore to allow the other boats to pass," reported Buchanan, "the Teche being here very narrow." The ironclad *Diana* and the tinclad *Kinsman* now took the lead, but the *Diana*'s gun misfired, leaving the *Kinsman* to advance alone toward the *Cotton*. Rebel field artillery appeared on the banks, blasting away at the solitary Union vessel. Battered by projectiles and taking on water, the *Kinsman* retreated down the Teche just as Buchanan reappeared with the mended *Calhoun*. "By running my bow into the bank," Buchanan recorded, "I brought my port broadside to bear on the *Cotton*." Sailors aboard the *Cotton*, having expended the last of their gunpowder bags, fired off a few more shots by improvising powder bags from the legs of their trousers. "She stood for about twenty minutes," Buchanan observed of the resolute *Cotton*, "when she backed up around a turn in the Teche, and soon got out of our range."[5]

After the *Cotton* departed, the Rebel artillery fell back, allowing Buchanan to grapple unmolested with the obstructions in the bayou. He made no progress, however, recounting, "As night was coming on I did not think it prudent to lay in the Teche, where the enemy could come all around us at night, and fire upon us with musketry and artillery without our being able to see them. . . ." The Union flotilla withdrew to Brashear City to make repairs and bury its dead. The *Kinsman* had endured the hottest fire that day: slugged by fifty-seven rounds (three piercing its flag), it counted four wounded and two dead.[6]

So began the Civil War on the Teche.

In mid-January Buchanan's gunboats returned to the bayou accompanied by infantry, artillery, and cavalry under General Godfrey Weitzel. These land forces hailed from Maine, Massachusetts, Vermont, Connecticut, New York, and Michigan. Many northern soldiers regarded the swampy semitropical region with a mixture of admiration and bewilderment. A New York officer recalled, "[W]e were reminded of one vast flower garden. The woods, beautiful with grey moss, which adorns every tree; the magnolia in full bloom—the thousands of wild roses, which spread our pathway on either side—seemed like the fairy-land of some olden tale." Others viewed the region less romantically. "[T]he swamp country of Louisiana," quipped one Massachusetts soldier, "seems at least one whole geologic age behind the rest of the world."[7]

Landed at the mouth of the Teche, these Union troops drove a Rebel garrison from Pattersonville, scouted the bayou west of town, and located their quarry. The "terror of the Teche," as one Union soldier called the *Cotton*, floated as before above the obstructions at Cornay's Bridge. The next morning the *Kinsman*, *Estrella*, *Calhoun*, and *Diana* steamed upriver to again confront the Rebel vessel. This time, however, Weitzel's land troops would shield Buchanan's flotilla, preventing a repeat of November's fiasco: on the Teche's east bank marched a regiment of Union infantry; on its west bank advanced the rest of Weitzel's forces, including four artillery batteries.

Buchanan, however, unwisely spurred his vessels ahead of their land escorts. He soon regretted his haste. "Directly a torpedo [mine] exploded under the stern of the *Kinsman*," recounted a Union sailor, "unshipping her rudder, and a sharpshooter from a rifle pit on shore shot [the ship's master] in the shoulder." Crippled by this turn of

African American soldiers of the northern Corps d'Afrique remove Rebel obstacles at Cornay's Bridge on Bayou Teche, September 1863. Source: *Frank Leslie's Illustrated Newspaper* XVII (November 7, 1863).

events, the *Kinsman* yielded the vanguard to the *Estrella*. But its timid commander refused to press forward, stirring Buchanan on the *Calhoun* to bark, "Then move out of the way and I will go!"

Oblivious to the barrage of enemy fire, Buchanan sprang onto the *Calhoun*'s deck to turn his spyglass on the assault. Moments later "Buchanan exclaimed, 'Oh, God!'" recounted one of his sailors. "The glass went over his shoulder. . . . He fell like an ox, and as he fell I saw a blood spot the size of a half dollar in front of his right ear and from it the blood began to flow." Now leaderless, the *Calhoun* ran ashore under a swarm of Rebel bullets that bounced off iron fittings and splashed in the dark waters of the bayou. As shipmates fell dead or wounded on all sides, the panicked crew reversed engines only to plow into the opposite bank. At last a fortuitous current whisked the damaged vessel downstream to safety.

Now the *Diana* moved to the fore, pausing to debark a unit of sharpshooters, who flushed Rebels from concealed rifle pits along the bayou. The *Cotton*, meanwhile, withdrew upstream, but Weitzel's sharpshooters caught up with the vessel, killing its second-in-command and maiming its captain, Edward Fuller. Despite severe injuries—"shot through both arms, the purple tide of life gushing from his wounds," waxed a Rebel newspaper—Fuller used his feet to steer the *Cotton* upstream to

Wreck of the gunboat CSS *Cotton* in Bayou Teche near present-day Calumet. Note the person standing amidships. Source: *Frank Leslie's Illustrated Newspaper* XVII (November 14, 1863).

safety. After his evacuation, the *Cotton* steamed back down the Teche to renew the contest. By then Rebel troops had disengaged the enemy, which also fell back for the day, intent on renewing its pursuit of the *Cotton* the next morning.

Pursuit would be unnecessary: Taylor's able lieutenant, General Alfred Mouton, decided to sacrifice the *Cotton* to impede a more formidable Union advance. That night Rebel sailors guided the doomed vessel on a "funeral trip down the bayou," as its commander recorded with regret, sinking the gunboat at an angle to block the entire width of the Teche. As a final measure the Rebels burned the vessel to the waterline. "[O]ne stroke of the mighty pen has swept into annihilation what tempests of shot and shell and fire had failed to scathe," lamented a southern journalist. "She now lies a gloomy wreck upon the water, though lost and abandoned, defiant in her loneliness, and still, as she was when afloat, a barrier to the advance of her foe." Content that their objective—sinking the *Cotton*—had been achieved, Union forces withdrew from the Teche and returned to their base at Brashear City.[8]

In late March the *Diana*, commanded by Acting Master Thomas Peterson, pulled two barges into the Lower Atchafalaya River to commandeer the sugar crop of a certain Widow Cochrane. Despite the widow's protests, Union troops grappled with her hogsheads of raw sugar, stopping only when Rebel cavalry appeared on the plantation's edge. Sensing a trap, Peterson withdrew to Brashear City and reported the incident to Weitzel.

In response, Weitzel sent the *Diana* back across Berwick Bay with infantry and cavalry. Once ashore, these land forces pursued the Rebel horsemen to the mouth of the Teche. The sight of enemy reinforcements, however, prompted the northern troops to return to the *Diana*, which ferried the expedition back to the opposite shore. The next day Weitzel again sent out the *Diana*, this time laden with nearly seventy New York and Connecticut sharpshooters. He ordered Peterson to steam to the mouth of the Teche via the Lower Atchafalaya inlet from Grand Lake (which extended farther south than it does today)—but to go no farther.

When the *Diana* reached the Teche, Peterson—boasting to a fellow officer that he could "blow any six of their batteries to pieces"—defied Weitzel's instructions. Steaming past the bayou's mouth and down the Lower Atchafalaya, he drove the *Diana* into a gauntlet of concealed Rebel cavalry and artillery. The ambush caught the vessel in a merciless crossfire of bullets, solid shot, and exploding shells. Peterson himself was mortally wounded, crying out as he fell, "Great God, they have killed me!" After three hours of combat that wiped out nearly all hands on the *Diana*, the Rebels seized the vessel and hauled it up the Teche for repairs: it was now a Confederate gunboat.

Union general Banks now resolved to smash Taylor's forces along the lower Teche. He would do so not only to disperse the Rebels who had bedeviled his gunboats, but, more importantly, to cut vital supply lines stretching from Texas to the Mississippi. To his staff Banks confidently stated that "he would make this campaign and then go home." To Union general Ulysses S. Grant he reported, "We shall move upon the Bayou Teche . . . , probably encounter the enemy at Pattersonville, and hope to move without delay upon [New] Iberia to destroy the salt works [at Petite Anse Island], and then upon Opelousas." Once there, Banks assured Grant, he would swing his forces east and attack Port Hudson, the Rebel stronghold on the Mississippi.[9]

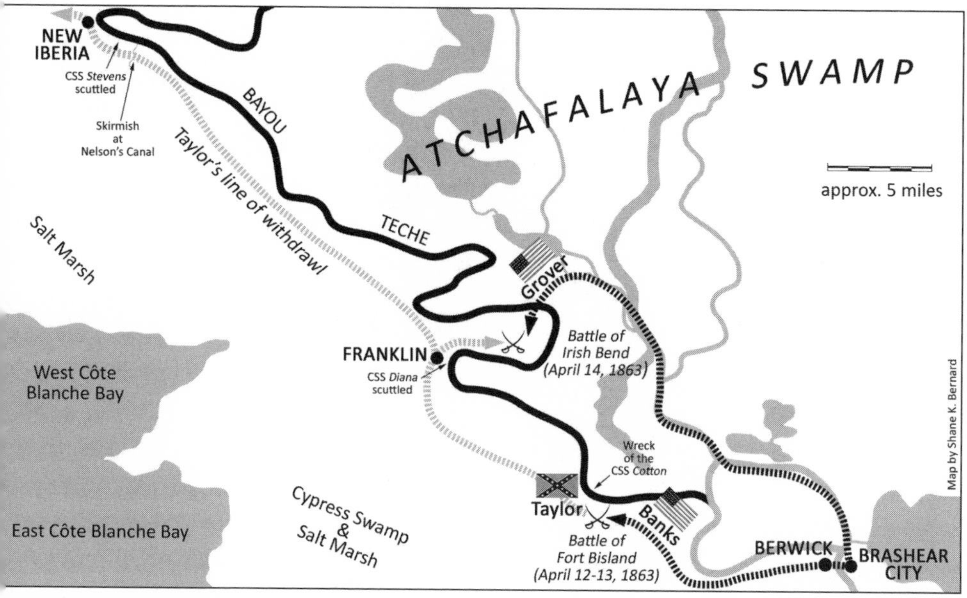

Civil War actions on the Teche, April 1863.

On April 9 Banks's troops crossed Berwick Bay aboard the familiar *Calhoun* and *Estrella* and two newly arrived gunboats, the *Clifton* and *Arizona*. In addition to these armored vessels and their crews, Union forces consisted of infantry, artillery, and cavalry from Maine, Massachusetts, Connecticut, New Hampshire, Vermont, New York, and Wisconsin. Numbering some ten thousand men, Banks called his force "the Army of the Gulf." He marshalled this army for two days at Berwick before advancing along the Lower Atchafalaya toward the Teche's mouth. His vanguard skirmished with Rebel cavalry near Pattersonville, captured the town, and next morning followed the Teche upstream.

After about a six-mile march, Union troops encountered a roughly two-mile-wide Rebel entrenchment. This formidable barrier ran southeast from a nearly impenetrable swamp to the east bank of the Teche. On the opposite bank the entrenchment crossed more cane fields to end in another swamp. Between the swamps and the banks of the bayou towered stalks of sugarcane, spoiling in the field because of wartime chaos.

On the far left of the Rebel line stood Redoubt Number 2, an earth-and-log stronghold protected by a moat. A half-mile behind the line's

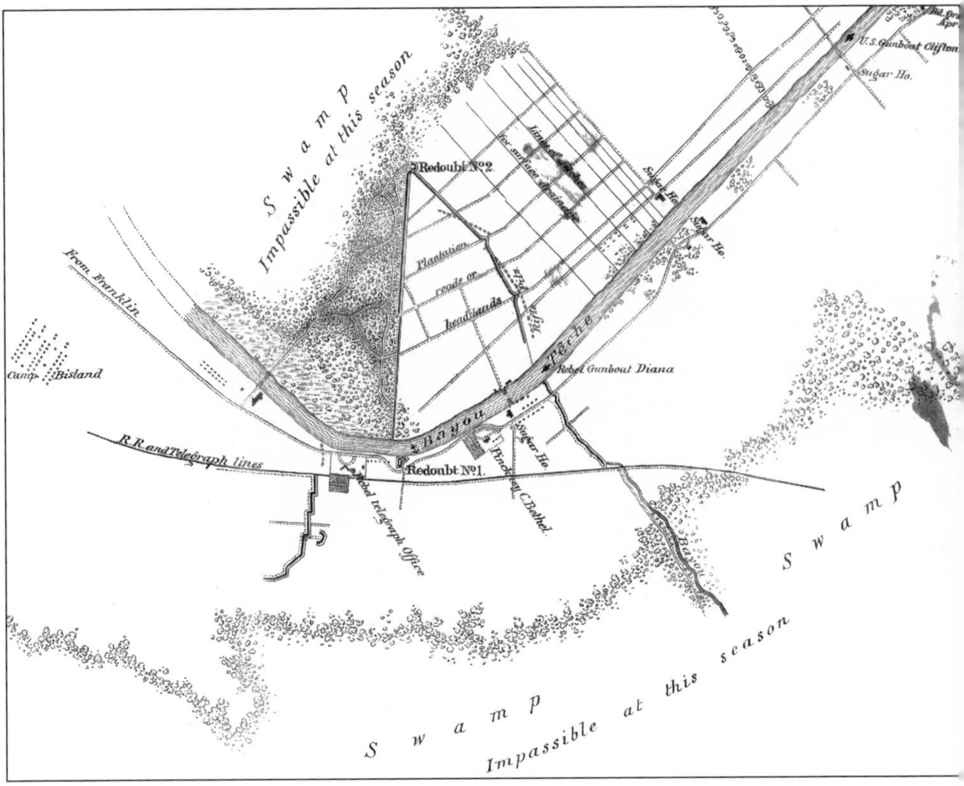

Union map of the site of the Battle of Ft. Bisland, 1863, fought on both sites of the Teche near present-day Calumet. "Redoubt No. 1" is the fort. Modern canal, levee, and highway projects have greatly altered the battlefield's topography, though part of it remains, as during the Civil War, in sugarcane. Source: Library of Congress, Washington, DC.

center loomed another earthwork. Sitting directly atop the Teche's west bank, this structure, called Redoubt Number 1, otherwise bore the name Fort Bisland (in homage to a local sugar planter). Next to the fort, on the Teche itself, floated the captured *Diana*, her cannons now trained on approaching Union forces.

Two features thus divided the landscape at Fort Bisland: the entrenchment and Bayou Teche. Both would heavily affect the conduct of the pending battle.

About four thousand Rebel soldiers—infantry and artillery reinforced by cavalry—manned the defensive line. Some, such as the St.

Mary's Cannoneers, the 10th Yellow Jacket Battalion, and the 18th Louisiana "Creole" Regiment, came wholly or partly from the Teche region. Others hailed from elsewhere in Louisiana, and from Texas and Arizona. Taylor called his small, ragtag force "the Army of Western Louisiana."

As historian Donald S. Frazier has observed, "Both sides were now in place, and the next few days might determine the control of western Louisiana, and with it, the Mississippi River."[10]

On April 12 Banks ordered a general advance on both sides of the Teche. Union troops groped their way through the sugarcane until the stalks gave way to "a long and comparatively narrow plain," as a Union officer observed. Across that flatland, about one-third of a mile away, spread the Rebel entrenchment, "barely visible to the naked eye." The scene, noted the officer, "was one of perfect quietness and silence and desertion."

Rebel cannons then opened fire, including those aboard the *Diana*—"our former staunch little gunboat," reflected a Union soldier, "with a large rebel flag flying." As a Vermont officer recorded, "The air was full of deadly missiles of every description—shells, solid shot, grape, and even pieces of railroad iron; and the earth was plowed in every direction as the huge projectiles buried themselves in the ground, throwing the dust and dirt over the men." Spotting the placement of the Rebel cannons, Banks ordered his infantry to take cover in the furrowed cane fields while his own guns returned fire. An hours-long artillery duel ensued, ending only at sundown.

That night Union engineers pilfered sugar coolers to erect a makeshift pontoon bridge across the Teche. Banks had previously used the captured steamboat *A. B. Seger* to move troops to the bayou's east bank, where Taylor's men under Mouton held the Rebel line. The pontoon bridge would now allow Banks to rapidly transfer men across the Teche. The Rebels, however, set up their own pontoon bridge, which spanned the bayou slightly upstream of Fort Bisland.

The all-out artillery duel continued in the morning while Banks prepared a frontal assault. The *Diana*, meanwhile, targeted the Union pontoon bridge: it missed, but struck closely enough to splash enemy troops with bayou water as they scurried across. Soon the improvised bridge listed steeply, sending a rumbling ammunition caisson into the Teche.

Responding to this constant gunboat harassment, Union batteries focused ten heavy cannons on the *Diana*. Projectiles burst through its armored shell, exploding in its cramped engine room. Taylor hurried to the edge of the Teche to inspect the gunboat so vital to his defense. "She was lying against the bank under a severe fire," recalled the general:

> The waters of the bayou seemed to be boiling like a kettle. An officer came to the side of the boat to speak to me, but before he could open his mouth a shell struck him, and he disappeared as suddenly as Harlequin in a pantomime. Semmes [commander of the *Diana*] then reported his [vessel's] condition. Conical shells from the enemy's Parrotts [rifled cannons] had pierced the railway iron, killed and wounded several of his gunners and crew, and cut a steam pipe. Fortunately, he had kept down his fires, or escaping steam would have driven everyone from the boat. It was necessary to take her out of fire for repairs. To lose even temporarily our best gun, the thirty-pounder, was hard, but there was no help for it.

Later Banks ordered a second advance on both sides of the Teche. Union infantry maneuvered in front of the Rebel entrenchment, threatening to storm it here and there, but again the battle devolved into little more than a clash of artillery. The event, remembered a Union officer, "was an artillery duel . . . with a dash of infantry charging and heavy musketry on either flank, and a dribble of skirmishing along the whole line."

That night Taylor stealthily withdrew his forces up the Teche toward Franklin, bringing an end to the Battle of Fort Bisland—with good reason. Rebels had discovered Banks's ruse: a flotilla of Union gunboats, transports, flatboats, and barges carrying horses, cannons, supply wagons, and about eight thousand soldiers under General Cuvier Grover had crept up Grand Lake to attack from behind. If successful, the maneuver would ensnare and crush the Army of Western Louisiana.[11]

Taylor sent a detachment to shadow Grover, whose amphibious assault came ashore around dawn. Once landed, Grover's troops marched toward a stretch of the Teche that looped back on itself to form a roughly twelve-mile oxbow. Locals called this oxbow Irish Bend.

That day, April 13, witnessed a running battle for several key bridges over the Teche. On the oxbow's upper reaches Rebels set alight the

'he Battle of Irish Bend, fought on an oxbow formed by the Teche near Franklin, April 14, 1863. Source: *Harper's* *Veekly* VII (May 16, 1863).

McWilliams plantation bridge; its destruction would seriously impede Grover's advance. Union forces pushed forward, however, and rescued the structure. Rebels then grabbed three other bridges—at the Porter, Bethel, and Simon plantations—and set them ablaze. Grover's men retook the former two bridges and doused the fires, only to torch the Bethel bridge later in the day to prevent its recapture.

Ultimately, the Rebels managed to demolish only the Simon bridge. But they successfully lured Grover's entire force down a bayouside road leading toward Franklin. Had Grover instead ordered some of his troops in the opposite direction, he would have found a shortcut bypassing Irish Bend. Known as Harding Lane, it would soon rumble with Taylor's main army, furtively retreating from the frontline at Bisland and out of Banks's trap.

By dawn on April 14 Taylor himself had arrived at Irish Bend. To his surprise Grover's force had encamped without securing Nerson's

Woods, the oxbow's most strategically important feature. "It was a wonderful chance," remarked Taylor. "Grover had stopped just short of the prize. Thirty minutes would have given him the wood and [the Harding Lane] bridge, closing the trap on my force." Taylor ordered his troops to occupy the woods before Grover realized his blunder. Pushing into the overgrowth, Rebels reached the far edge of Nerson's Woods to discern sugarcane fields and, beyond them, groggy enemy troops heading down the Teche road toward Franklin—still unaware of the crucial shortcut that lay behind them.

Grover's troops detected the Rebels in Nerson's Woods, but wrongly thought them only a handful of pickets. Veering right, the northerners crossed a cane field to address this nuisance. "As we swung into position," recalled a Union soldier, "we suddenly heard the cry, 'Attention Battalion, take aim, fire!' and immediately the woods seemed to spring into life, while a perfect storm of canister, grape and minie balls was rained down upon our ranks."

For the next three hours Union and Rebel infantry clashed along the tree line of Nerson's Woods, joined to a lesser extent by artillery and (on the Rebel side) dismounted cavalry. Pinning down Grover's troops, the Rebels sensed the moment for a charge. "[S]uddenly there was a terrific yell," observed a Union officer, "and 1,100 men rushed in on our flank and commenced peppering us well. I have heard men speak of a hailstorm of bullets; but I never imagined it before. . . . [T]he balls whisked and zipped among the cane-stalks and ploughed up the ground around us. . . . In less than ten minutes two-thirds of all the loss we experienced on that day occurred. . . ."

Poised to annihilate the enemy at Irish Bend, the Rebels suffered two setbacks in quick succession: a bullet mortally wounded the Confederate officer leading the charge, and a Union salient along the bayou seemed about to fall on them from behind. Leaderless and confused, the Rebels retreated. Grover then succumbed to over-caution; he consolidated the Union line, but declined to renew the attack. Taylor, meanwhile, regrouped his forces, newly swollen with reinforcements. Moreover, the *Diana* arrived from Bisland to lob shot and shell at Grover's troops.

Soon the last of Taylor's main army passed up Harding Lane to safety. The Rebels engaged at Irish Bend withdrew in the same direction, while the *Diana* continued to distract Grover's troops. The fighting was over, and the Union army held both battlefields. Yet Taylor had out-generaled Banks at Bisland and Grover at Irish Bend, dividing his command and pulling forces out of combat to fight another day. And he did so while inflicting losses on a superior foe, particularly in the battle on the oxbow. "[T]he fight at Irish Bend," grumbled a northern officer, "was a needless murder of our troops."

"Toward night," recorded a Union corporal of the aftermath at Irish Bend, "I go down the cart-path to the actual field, and see the broken muskets, the scattered knapsacks and clothing, the furrows where the enemy lay, the bloody pools where the dying fainted, the burial parties, and the piles of distorted corpses lying by the trenches just dug to receive them." A sugarhouse on the edge of the Teche became a field hospital; it witnessed as bloody a scene as any found at other, more commemorated Civil War battles. "The surgeons . . . ," observed one Union soldier, "make a pile of legs and arms, feet and hands. . . ."

"Our little battle is known among the men as 'Irish Bend,'" a veteran reminisced two decades later. "It does not make much of a figure in history because only a division was present and not all of that fully engaged, but it was sharp, obstinate, and bloody, was skillfully handled, and was as truly a battle as Gettysburg or Shiloh."[12]

As the last Rebels withdrew, they burned their unarmed supply steamboats *Newsboy*, *Gossamer*, and *Era No. 2* on the Teche at Franklin. Taylor also ordered the destruction of the *Diana*. When flames reached its store of gunpowder, the ironclad exploded in "a stunning, deafening crash," as a Union physician noted. "The earth quivered with the violence of the concussion," he continued, "and the air was filled with a sulphurous cloud, and flying sticks and timbers. . . . The gunboat *Diana* was no more."

The Confederate supply boat *Cornie*, also moored at Franklin, escaped destruction, if only because it served as a floating hospital. When enemy troops entered town, the *Cornie*'s master surrendered the vessel to a wounded Union prisoner. "Take charge of her, sir," he directed the northerner, "and hoist your flag on her."

Farther up the Teche, Taylor dispatched troops to scuttle the unfinished Rebel gunboat *Stevens* (formerly the *Hart*), which they sank between New Iberia and Jeanerette to block encroaching Union vessels. He also ordered the supply steamers *Darby*, *Uncle Tommy*, *Louise*, *Blue Hammock*, and *Crocket* (sometimes identified as the *Cricket*) to evacuate the Rebel arsenal at New Iberia. Once loaded, they scurried toward the safety of the upper Teche. Taylor then marched his troops away from the bayou, moving northwest toward Vermilionville (now Lafayette) and, beyond it, Opelousas and Alexandria.[13]

Back at Irish Bend, Banks's command arrived from Fort Bisland to learn that the clever trap set for Taylor had failed. Incensed, but knowing the exhaustion of his troops, Banks delayed pursuit of the fleeing Rebels until next morning. When his army finally advanced, the gunboat *Clifton* provided a reassuring escort. "We bathed in the Teche," recorded a Union corporal, "and watched the *Clifton*, our old, grim friend, which came steaming up. . . . [T]his gunboat in the bayou . . . was to guard the flank of the pursuing column." A northern lieutenant described "the strangest sight I have ever seen . . . the *Clifton* plowing her way up the stream, her sides almost touching the banks. . . . The ground is a good deal higher than the stream and [from] a little way off the boat looked like a mammoth machine moving on the surface of the ground." Below New Iberia, however, the remains of the *Stevens* halted the *Clifton*'s ascent.[14]

After occupying New Iberia, Banks sent a detachment farther up the Teche to search for Taylor's supply steamers. Another detail seized the Rebel salt works at Petite Anse Island. Most of his forces, however, veered northwest and continued to pursue the Army of Western Louisiana, now at Vermilionville.[15]

The Union detachment chasing the steamers marched through St. Martinsville to the strains of "Yankee Doodle." Later that day it reached the village of Breaux Bridge. There it found the so-named bridge over the Teche in flames. Resuming its march upstream after a two-day delay, the unit arrived at Arnaudville to discover the sought-after Rebel steamboats—torched because the vessels could go no farther up the increasingly shallow bayou. "I learned that the *Darby* had three guns on board when she was burned," reported the unit's commander, "and that they could be seen above water when the fire had died out":

> Before leaving the junction [of bayous Teche and Fuselier at Arnaudville] I examined the wreck of four steamboats. The water having risen after the rains . . . I was unable to see any names on the boats, or the guns reported to have been left on the *Darby*. Their smoke-stacks and part of their machinery only were above water. From all the information I received I have no doubt of their being the *Darby*, *Louise*, *Blue Hammock*, and the *Crocket*. The *Uncle Tommy* is burned higher up in the Bayou Teche, and the wreck of this boat is high enough out of water to see her name. Cargoes of beef, rum, sugar, and commissary stores, cloth, uniforms, and large quantities of arms and ammunition were destroyed in these boats. Some barges took off portions of the cargoes of ammunition and arms from these steamboats before they were set on fire. I passed the charred remains of two or three barges in the Bayou Teche and I found by the roadside empty musket boxes and large quantities of shot, salt, some new cavalry sabers, and a few barrels of oil. I think there must be a valuable portion of the cargoes of the destroyed steamboats in the houses and woods along the banks of the Teche and in the vicinity of the junction, and possibly [on] some barge hidden in the Teche. . . .[16]

After examining the wreckage, the Union detachment rejoined the main body of Banks's army, which advanced through Opelousas to take Alexandria in central Louisiana. There the first Teche campaign fizzled out: too timid to chase Taylor deeper into Louisiana's interior, Banks swiveled his army east, crossed the Atchafalaya swamp and Mississippi River, and besieged Port Hudson. Within two months the Rebels would retake all of Bayou Teche down to its mouth, and would even cross Berwick Bay to capture Brashear City.

General Banks and his Army of the Gulf would return to the Teche Country twice more. Unlike the first Teche campaign, only minor skirmishes occurred along the bayou during those invasions. Yet all three campaigns shared two salient features: the mass exodus of enslaved blacks in search of freedom; and widespread looting of civilian property by the Union army, punctuated by sporadic acts of arson. "As the enemy . . . retreated he used the torch," observed an official Confederate report, "leaving some evidence of [his] presence, in chimney stacks arising out of the charred ruins of costly edifices. These still stand, marking the places where once stood the elegant and hospitable mansions of [Teche sugar planting families] . . . the Stirlings, the Wilcoxons [Wilcoxsons], the Fusiliers, the Carpenters [Charpentiers], the Corneys [Cornays], the Perkins, the Bethels . . . the Hardings, the Burns [Byrnes], and others."[17]

Union soldiers broke into houses, seized valuables at gunpoint, and rejoined the march laden with plunder. "One who has never seen a house 'sacked' by the 'boys,'" observed a Union officer near New Iberia, "can have no idea how faithfully they 'do their work.'" One of Banks's own generals condemned the practice as "disgusting" and "disgraceful," recording in an official dispatch: "Houses were entered and all in them destroyed in the most wanton manner. Ladies were frightened into delivering their jewels . . . by threats of violence toward their husbands. Negro women were ravished in the presence of white women and children."

Some Union commanders tacitly encouraged and even participated in this depredation. At the end of the first Teche campaign, for example, Colonel Thomas E. Chickering gathered up the spoils of the countryside, heaped it into overflowing carts at Port Barre, and escorted it down the banks of the Teche to the federal depot at Brashear City. The ramshackle wagon train stretched for eight miles and included fifteen hundred cattle, two thousand horses, and five thousand "contrabands" (liberated slaves). "[T]heir slaves were all gone," recalled a freedman formerly enslaved to the Wyche family, whose Belmont plantation overlooked the Teche near New Iberia. "The barns and houses had been burned. Vegetation had been trampled down. Chickens, mules, cattle, all had been driven off. There was not a living thing on the countryside except a few faithful slaves and the women folks. The army had moved on."[18]

When in spring 1865 the war ended with the South's defeat, the Teche lay choked with burned bridges and sunken vessels. Four years of wartime neglect further obstructed the bayou with rotting stumps, submerged logs, and oozing mud bars. Moreover, the brutal clashes along its banks had destroyed not only human lives but livestock, dwellings, sugar houses, fences, wagons, and the vital cotton and sugar crops. Years would pass before the Teche Country recovered, and in some ways it never did.

5

Hard Times on the Bayou

Teche Country rebels tramped home after the Civil War to resume their fractured civilian lives. Back on the bayou, however, they found the region devastated by its repeated Union occupation. Worse, many returned too late in the year to plant essential crops and had no money to buy provisions from New Orleans. "Most veterans," wrote historian Conrad, "therefore resorted to growing a few vegetables, scaring up a few chickens, a coon, or rabbit while waiting for spring to come." For a time matters seemed hopeful. "Fences are springing up, repairs are being made, plows are moving," a spectator optimistically noted from the bayou in early 1866. But it was actually the beginning of hard times on the Teche. Sugar and cotton prices sank precipitously, while the federal program called Reconstruction imposed major—some would say severe—changes on the Teche. Among these changes, civil rights for freedmen and the punitive disenfranchisement of former Confederates sparked an intense backlash, exacerbating the era's pervasive violence, lawlessness, and racial conflict. Strained already by these tumultuous events, the postbellum Teche also experienced recurrent outbreaks of disease and a series of calamitous floods that dragged the region even deeper into anguish. "Poverty gripped the Teche Country" noted Conrad, ". . . and people went hungry."[1]

Many of the bayou's planter elites struggled without success to recover their antebellum wealth. In the late 1860s and early 1870s, for example, St. Mary, St. Martin, and newly formed Iberia Parish together made only half as much sugar as St. Mary alone before the conflict. The local sugar industry's decline stemmed from the shift from slave to wage labor; the disrepair of field and mill equipment; and a lack of seed cane arising from the wartime neglect of sugar fields. Discouraged by these hindrances, sugar planters on the lower Teche sought relief by emulating those on the upper Teche: they tried to grow cotton, despite previous failures at large-scale cotton planting below the middle bayou region. Cotton, however, demanded less capital and fewer workers than sugar, and at the time fetched an enticingly higher price.[2]

Regardless, the lower Teche's partial shift to cotton did nothing to buoy the local economy. This failure sprang from a string of disasters that hit the bayou in rapid succession after the war. Drought and army worms destroyed the crop of 1866, and next year the insidious pests struck again. Moreover, cotton prices soon collapsed, losing about half their value between 1866 and 1870, and half their value again by 1876. Some planters along the Teche retreated from cotton, only to return a generation later, when a St. Mary Parish resident described a veritable "cotton 'craze'" among local planters. "Every day," he noted, "reports come in of planters who are going to try their hand at cotton raising." But as a journalist sardonically observed, "Teche parishes produced this last year a hundred thousand bales of cotton and of course have had to unload it onto a four-cent market." Planters who remained loyal to sugar likewise fared poorly. Overplanting and increased competition (particularly from Cuba) caused sugar to lose a quarter of its value between April 1866 and April 1872, as the price fell from sixteen to twelve cents per pound. By January 1878 sugar prices had fallen by nearly half. Prices would continue to decline into the next century, so much so that by 1910 the commodity had lost about 75 percent of its former postwar value and cost less than five cents per pound.[3]

Struck by these tumbling prices, many once affluent planters became mired in substantial debt. The global Panic of 1873 added to their despair, forcing New Orleans banks to halt loans and to call in their debts. Refused credit by these skittish banks, some planters were compelled to sell their plantations, including their ornate mansions, else face the humiliation of foreclosure. Northern corporations and entrepreneurs (vilified as "carpetbaggers") often stepped in to purchase these bankrupt plantations, a trend that had materialized almost as soon as the Civil War ended. Around 1869, for example, an observant steamboat captain noted the influx of "Northern capitalists" along the Teche, all of whom "have purchased [plantations] within the last four years." This financial crisis wounded not only the planter class but the businessmen, artisans, yeoman farmers, tenant farmers, and sharecroppers who also resided on the bayou. Many descended into a state of poverty that lasted for generations. Indeed, as historian Brasseaux has commented, there is strong evidence that rural south Louisianians, including those along the Teche, became so poor that in 1929 they did not notice the onset of the Great Depression.[4]

In the meantime, struggling Teche Country planters became so disillusioned with sugar and cotton that many dared to embrace large-scale rice farming, an industry entirely new to the region. Although rice previously had been grown on the bayou in a small way, planters now sought to cultivate the grain as a major cash crop. This strategy required them to overcome several challenges. To properly grow rice, for example, planters had to enclose their fields with levees and install expensive steam pumps, used to draw large quantities of irrigation water from the Teche. These pumps soon became common sights along the bayou. "In the ten miles between New Iberia and Jeanerette," a journalist recorded in the early twentieth century, "the writer counted eighteen rice pumps bringing the high waters of Bayou Teche to as many rice plantations of no small acreages."[5]

The effort was successful for a time. In 1869, on the cusp of the move toward rice, the entire Teche Country produced no more than 343 barrels of rice. Seventeen years later rice production reached 42,290 barrels in Iberia and St. Mary parishes alone. This trend would continue for decades, so that in 1914 a canoeist on the Teche observed "more rice fields than cane" between New Iberia and Charenton. Ultimately, however, this experiment in rice production, like that in planting cotton, proved less profitable than growing sugarcane. Teche rice faced stiff competition from not-too-distant Acadia Parish, whose vast, open prairies and high clay pans were better suited than the Teche Country to growing rice. (Although rice crops disappeared long ago from the banks of the Teche, the Conrad Rice Mill in New Iberia is a remnant of the area's rice-growing days.) Sugar, therefore, endured as the primary cash crop along the Teche, even in the face of chronically low prices bolstered by protective tariffs. The numerous old, family-owned sugar houses, however, fell into disrepair, replaced by a handful of modern, corporatized, and heavily mechanized mills located at strategic points along the bayou. Some of these operated as co-ops, built by sugar planters who pooled their capital and shared the mills during grinding seasons.[6]

Besides economic turbulence, the Teche Country, like much of the South, experienced intense Reconstruction-era violence, much of it racial in nature. This trend emerged from the wartime breakdown of moral standards and civil institutions, and involved reactionary Democrats grasping for political control over newly freed blacks and their radical Republican sponsors. From obscure backwoods lynchings to massive civil unrests like the Colfax Massacre, the Battle of Liberty Place in New Orleans, and, closer to the Teche, the Thibodaux Massacre, racial violence became commonplace in the state and erupted on the bayou itself. There, both planter elites and other whites regarded the region's larger, now ostensibly free black population with palpable dread.[7]

Many whites along the Teche embraced violence, or the threat of violence, to preserve their racial hegemony, and to reclaim Louisiana's political machinery. Although the Ku Klux Klan never gained a foothold in heavily Roman Catholic south Louisiana—the Klan was not only anti-black, but anti-Catholic—whites in the bayou country emulated the Klan by creating their own terror groups. These included the Knights of the White Camellia, founded along the Teche in Franklin in 1867. No fringe organization, the Knights attracted up to half the white population of the state's Cajun parishes. In the Teche parishes of St. Landry and St. Martin, according to historian Brasseaux, the group attracted "almost the entire white population." Seven years later whites in Opelousas, about four miles from the Teche, established an even more virulent racist group, the White League. Its goal, noted Brasseaux: "race warfare to secure the end of Louisiana's black political domination."[8]

These militant white supremacists targeted freedmen up and down the Teche. An 1869 inquiry by reformist state politicians, for example, uncovered several incidents of racial brutality along the bayou. As the inquiry reported:

In 1866 a colored man was found hanged at "Grande Pointe" [Cecilia]. It is still a mystery who committed the deed, and no effort was ever made to find the guilty party.

March 10th, of the same year, a fishing party, composed of a young man named Caytan, aged sixteen, a woman and two young girls were seated in a small boat fishing in the "Bayou Teche," in front of Onesiphore Delahoussay[e]'s plantation. The young man was at one end of the boat eating, when the report of a gun was heard, and he received a load of shot in his head from a gun in the hands of Onesiphore Delahoussay[e], Jr., who then had the whole party arrested on the pretense that they were stealing wood from his place. . . . The young man lost one eye as the result of this wound. His parents took the necessary steps to bring the case before the court, but lawyers refused to take charge, and judges refused to receive their affidavit . . . [and] no suit was brought because the constables refused to execute the warrant of arrest.

The same year [as the above] a colored man named "Jean Louis" was killed in the woods at "Fausse Pointe" [the Loreauville area], no effort was made by the authorities to find the criminal party, and the death of that man is still a mystery.

The same year a colored girl was employed as a servant by a family named Melancon; after several months service a difficulty arose between the girl and Mrs. Melancon, the latter refusing to pay the girl her wages. She brought a suit before the justice of the peace at St. Martinsville. . . . When she went back to Melancon's plantation for her clothes, [Mr.] Melancon seized her in a rage and whipped her unmercifully, asking this poor orphan girl if she did not know that she had no right to bring his wife, "a white lady," before the law? Thereupon the girl came back to make her complaint again before the authorities, but the judge refused to entertain her case further.

In 1867, at "Fausse Pointe," a colored man named "Jean Baptiste" was with a fishing party on the lake shore [probably Lake Fausse Pointe, east of Loreauville, or adjacent Lake Dauterive]. He had absented himself for a few moments from the party, when the report of a gun was heard, and when his friends reached the spot he was found dead. His murderer, a white man, was standing near him with a gun, claiming as a reason for the act that the Negro was stealing horses, which was false, and witnesses can testify to it.[9]

At times Reconstruction-era racial violence along the Teche occurred on a much larger scale, immersing entire communities in turmoil. In

1873, for example, French Creole attorney, former Confederate officer, and Knights of the White Camellia cofounder Alcibiades DeBlanc marshaled about five hundred militiamen and, according to some reports, a couple of cannons to besiege radical Republican officials in St. Martin Parish's courthouse. Federal authorities responded to this "insurrectionary war in St. Martinsville," as a state newspaper called it, by sending U.S. troops up the Teche to restore order.

Arrested, DeBlanc evaded prosecution and returned to St. Martin Parish to foment yet another uprising the next year. Allegedly armed with more cannons—scuttled Civil War relics pulled from the Teche—his partisans again occupied St. Martinsville. Raiding the countryside, they seized firearms from black residents and ran off several Republicans on pain of death. Within the month vigilantes lynched two black Iberia Parish residents—murders a deputy U.S marshal implicitly linked to DeBlanc's militia. (Three years later, in a gesture reflecting the injustice of the period, the state's first post-Reconstruction governor, Francis T. Nicholls, appointed DeBlanc an associate justice of the Louisiana Supreme Court.)[10]

Racial violence continued to occur along the Teche even after Reconstruction ended in 1877. The very next year, for example, vigilantes in Franklin abducted a certain Moustand, a black prisoner accused of attacking two white women. The mob hung him from a belfry, "cut his throat, and threw the body into Bayou Teche." In 1889 the body of a black man, Jim Rosemond, "was discovered swinging on the bridge which crosses the bayou" at New Iberia. A band of local "regulators" had recently ordered Rosemond, a suspect in several crimes, to leave Iberia Parish on penalty of death. Race fueled an 1884 riot in Loreauville that took the lives of sixteen blacks and two whites. Mob violence again broke out on the bayou in 1887, when vigilantes in Pattersonville murdered at least four striking black plantation workers. With their civil liberties quashed by terrorism and the advent of Jim Crow laws, blacks along the Teche, as elsewhere in "the bayou state" and the South, would remain an oppressed underclass well into the twentieth century.[11]

Already marred by economic distress and racial violence, the Teche Country also witnessed a resurgence of disease. Cholera, for example, struck the region in 1873 during a national epidemic. An eruption on

ountry store along the Teche during one of the bayou's postbellum floods. Source: *Harper's Weekly* XXVIII
rch, 29 1884).

Calumet Plantation that year resulted in forty sufferers, eight of whom died. As in the antebellum era, however, it was yellow fever that most troubled Teche Country residents. In summer 1867 the disease struck the region for the first time in over a decade. New Iberia "was ravaged," a local physician recorded, who called the outbreak "a fierce epidemic." Of the town's roughly eighteen hundred residents, as many as seven hundred contracted the illness. Of those, seventy died. The disease also visited Jeanerette and St. Martinsville that year, and struck down several members of the Chitimacha tribe at Charenton, where, according to one flippant journalist, "All who did not die left, so that finished the matter." Another journalist deemed the outbreak "terribly severe" at Loreauville. "Out of two hundred and ninety-six inhabitants," he continued, "ninety-four died in less than three months. . . ." The fever came again to the Teche in 1878, 1897, 1898, and 1905, infecting or killing hundreds. Scientists finally linked the disease's spread to mosquitoes and launched an aggressive mosquito-control campaign. After the 1905 outbreak, yellow fever epidemics never returned to the Teche or to any other place in the United States.[12]

Then there were the floods. In 1867 a crevasse (break) in the Mississippi River levee sent floodwaters raging across the Atchafalaya swamp to engulf the Teche's east bank. Additional floods came in 1874 and 1882, the latter ranking as the worst local disaster in memory. That deluge originated in nonstop Midwest rains, whose waters eventually burst south Louisiana levees holding back the Mississippi. Floodwaters surged down the Atchafalaya and the Courtableau into the Teche and also spilled into the bayou through old canals connecting it to Grand Lake. "Dwelling houses, stores, sugar houses, corn cribs, and all sorts of buildings were wrecked and washed away," lamented an agricultural journal. In the Loreauville area alone, it reported, "forty houses of various sorts floated from their places with their contents." Southward from Ricohoc plantation near Centreville, noted the *Daily Picayune*, "there is one vast sheet of water over nearly every inch of ground to the Gulf of Mexico."[13]

The Flood of 1927, however, would be the most destructive. It affected not only the Teche Country, but seven Southern and Midwestern states. In fact, most of the floodwater that overwhelmed the Teche that year came from torrential rains in states like Iowa, Illinois, Ohio, Indiana, and Kentucky. As the enormous runoff from these downpours flowed toward the Gulf of Mexico, it breached levees along the Mississippi River and numerous connecting waterways. One of those broken levees ran along Bayou des Glaises in Avoyelles Parish. Located about thirty miles due north of the Teche's headwaters at Port Barre, the crevasse at Bayou des Glaises at first seemed too distant to worry Teche Country residents. Another breach, however, opened on the Atchafalaya River at Melville, some sixteen miles northeast of the Teche's headwaters. These two rampant floods intermingled, combined into one enormous flood, and quickly moved southward over open farmland and through backwoods toward Port Barre.[14]

Meanwhile, floodwaters gushed from the Mississippi into the Atchafalaya River and its adjacent swamp basin, pushing against a small protection levee running between Port Barre and Butte La Rose.

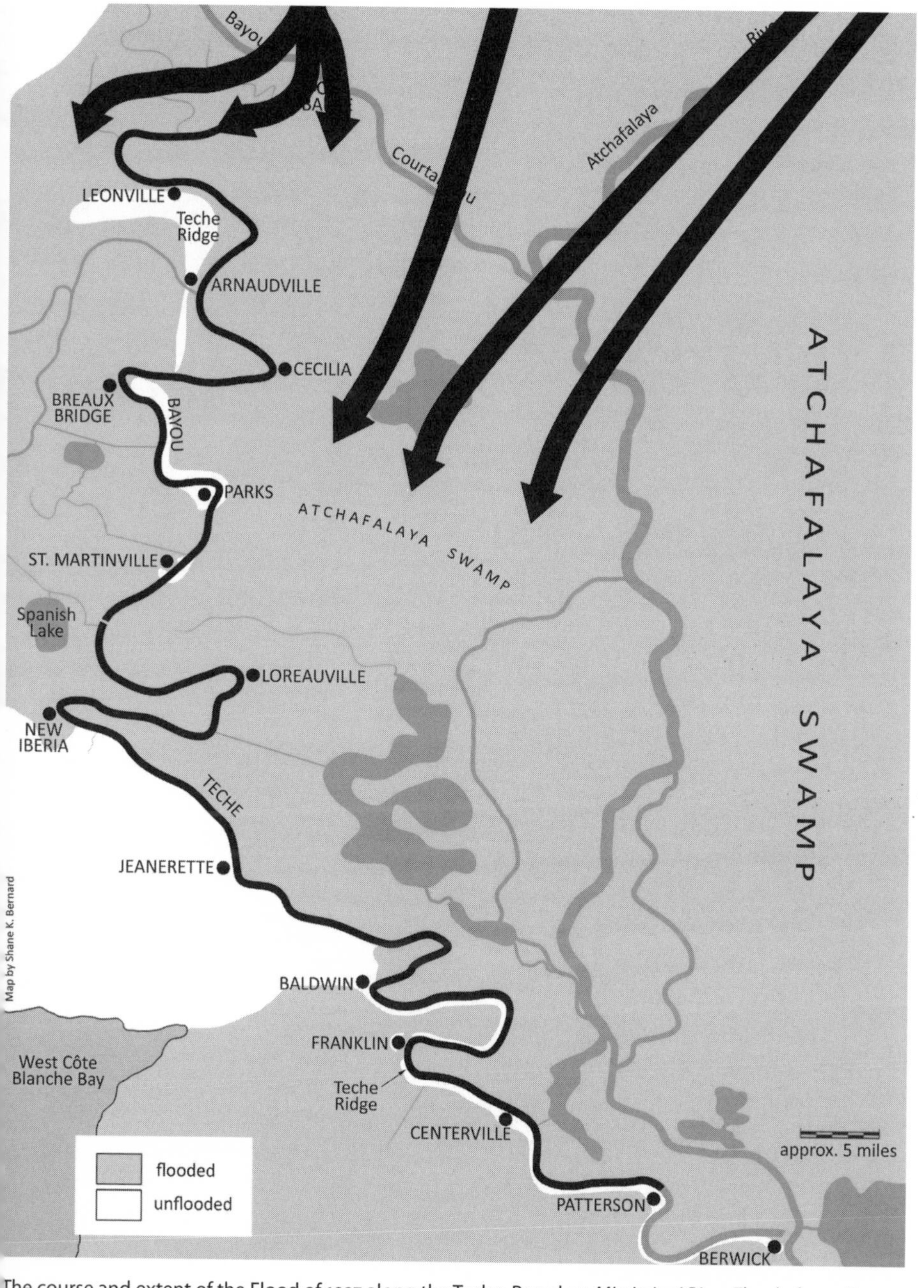

The course and extent of the Flood of 1927 along the Teche. Based on *Mississippi River Flood of 1927 Showing Flooded Areas and Field of Operations*, map (Washington: U.S. Coast and Geodetic Survey/U.S. Army Corps of Engineers, 1927).

Port Barre during the Flood of 1927. Source: L. V. Bernard Collection, in the author's possession.

Constructed as a safeguard against minor spring flooding, this levee burst at Cecilia and Henderson (near Breaux Bridge) and sent water rushing toward the Teche. Back upstream, the rising waters from Bayou des Glaises and Melville poured around both sides of the Teche and raced along the natural ridges forming its banks—toward Leonville, Arnaudville, Cecilia, Breaux Bridge, and other Teche Country communities. Thus, floodwaters ran down the east and west banks of the Teche, as well as down the Teche itself. "Towns on Bayou Teche in Grave Danger as Water Sweeps Southward," warned the U.S. Weather Bureau. "[R]esidents of this section should hold themselves ready for prompt retreat from the flood."[15]

Towns and rural districts along the bayou flooded one by one as the deluge moved downstream. Port Barre had been the first to flood, its streets filling with several inches of water by May 18. Soon the muddy surge reached the second story of some Port Barre structures. On May 20 the media reported, "Water in Arnaudville and Leonville is several feet deep and rising rapidly." Bridges in the former, a journalist noted, "are being washed away like chaff." A bridge near the junction of the Teche and Bayou Fuselier washed away seconds after "several score refugees" fled across the structure. The same journalist soon witnessed

he Evangeline Oak, St. Martinville, during the Flood of 1927. Source: L. V. Bernard Collection, in the author's ossession.

"residents of Breaux Bridge . . . hastily retreating in sight of the fast approaching waters."[16]

On May 22 the flood reached Breaux Bridge, then followed the Teche toward the parish seat at St. Martinville (by then using the modern spelling of its name). "Between Parks and St. Martinville," reported the *Times-Picayune*, "the road is rapidly going under water." Meanwhile, residents to the south paid no attention to warnings from the U.S. Weather Bureau and others. When the Bureau issued a flood warning for St. Martinville, locals "rose up in arms," recalled a meteorologist, "and declared they would not have flood waters, for they had never been flooded." By May 23, however, the waters had invaded St. Martinville. Across the Teche from downtown, the neighborhood of Pinaudville found itself under six feet of water. This refusal to heed warnings prompted visiting flood relief czar and U.S. Secretary of Commerce Herbert Hoover—who next year would become President of the United States—to declare, "People along the lower Teche ridge are moving slowly . . . thinking that the waters will not reach them. . . . The time for Bayou Teche residents, on both sides of the stream, to move is now."[17]

Around Loreauville cowboy volunteers from east Texas rounded up about 500 head of cattle for transport to higher ground. Parts of the town would soon be under at least four feet of water. Residents there sought refuge downstream in New Iberia and Jeanerette—an unwise decision because those towns, like Loreauville itself, sat directly in the flood's path. "Panic-Stricken Thousands Flee before Flood in Teche," declared a *Times-Picayune* headline. Another proclaimed, "Torrent Raging in Teche Moves on New Iberia."[18]

On the west side of the bayou the deluge—funneled between the Teche Ridge and the long escarpment to the west known as the *coteau* (slope)—poured into Spanish Lake, filled and overflowed the lake basin, then continued southeast toward New Iberia. Around 2 a.m. on May 26 the U.S. Weather Bureau received a frantic call from New Iberia's mayor. "[T]here is a wave of water now about five miles above here and coming down this way," spoke the mayor. "What is it going to do?" The town would flood, replied the Bureau, just as it had advised him days earlier—a warning the mayor had greeted at the time with skepticism.[19]

By 9:30 a.m. the flood reached the edge of New Iberia, coming at first not from the roiling Teche, but from Spanish Lake. Floodwaters also arrived across the Teche, swamping the east-bank neighborhoods opposite downtown New Iberia. Ultimately, downtown would receive up to three feet of water; the east bank, however, received up to six feet. (A hand-carved mark in the brick façade of the Lutzenberger Foundry, which sits on the west bank of the Teche near downtown, shows a flood-water depth of 4.75 feet; another mark inside the building, whose floor sits below street level, records a flood-water depth of 7.25 feet.)

The downstream communities of Jeanerette, Baldwin, Franklin, Centerville, and Patterson (the latter two, like St. Martinville, having modernized their names in recent decades) escaped the flood largely unscathed. Franklin, for example, saw flooding directly along the bayou, but otherwise remained dry. This occurred partly because it and the other lower-Teche towns sat on the Teche Ridge and also because the terrain below New Iberia opened up suddenly, allowing the floodwaters to spread out, run shallower, and dissipate their strength before emptying at last into the Gulf of Mexico. Regardless, most low-lying rural districts near these Teche-side towns flooded, as did Charenton, whose streets filled with several feet of water moving "with great force" toward Grand Lake.[20]

At the disaster's peak over half of Iberia and St. Martin parishes, more than two-thirds of St. Landry Parish, and approximately 90 percent of St. Mary Parish lay underwater. The number of displaced persons in these parishes reached about 90,000. Racially segregated refugee camps sprang up in bordering dry areas, feeding, housing, and treating flood victims until they could return home. There they found their dwellings and crops encrusted with mud, their farm animals dead, and their fresh-water wells contaminated with bacteria that, in the midst of so much misery, sparked a typhoid outbreak. "[N]o power on earth could have saved us from inundation by the greatest of all floods since the days of Noah," a Teche newspaper remarked at the time with forgivable hyperbole. At the urging of Teche Country residents, however, federal, state, and local governments would try to prevent future floods. In doing so, they would forever change the character of the Teche.[21]

6

Designing the Bayou

While the Teche Country suffered under the combined weight of poverty, disease, violence, and flooding, some residents voiced concern about the miserable condition of Bayou Teche. "The almost complete obstruction of this stream is a great drawback on commerce and agriculture," declared steamboat captain E. B. Trinidad around 1868. A leading Teche booster, Trinidad pushed government officials to allocate tens of thousands of dollars for the bayou's improvement. To win support he printed up a crude map (an original of which now resides in the National Archives) detailing the Teche's many navigational hazards: not only snags, rafts, and sand bars, but sunken vessels—including four schooners, eleven steamboats, three Confederate transports, and five Confederate gunboats. (In fact, one of the five was not a gunboat, and another he counted twice because it had broken in two.) Trinidad's map also documented the wreckage of a Confederate "mud-digger" (dredge boat) and "torpedo machine" (minelayer)—not to mention a sunken pontoon bridge, possibly one of those used in the Battle of Fort Bisland.[1]

Because of these complaints by Trinidad and others who depended on the waterway for their livelihoods, the state of Louisiana set aside $30,000 in 1869 "for removing obstructions in navigation in the Bayou Teche." On completion of this project, however, Trinidad expressed only displeasure, charging that "personal greed absorbed all of the state appropriation. . . ." He possibly felt relief when in 1870 the U.S. Army Corps of Engineers took over clearing the Teche.

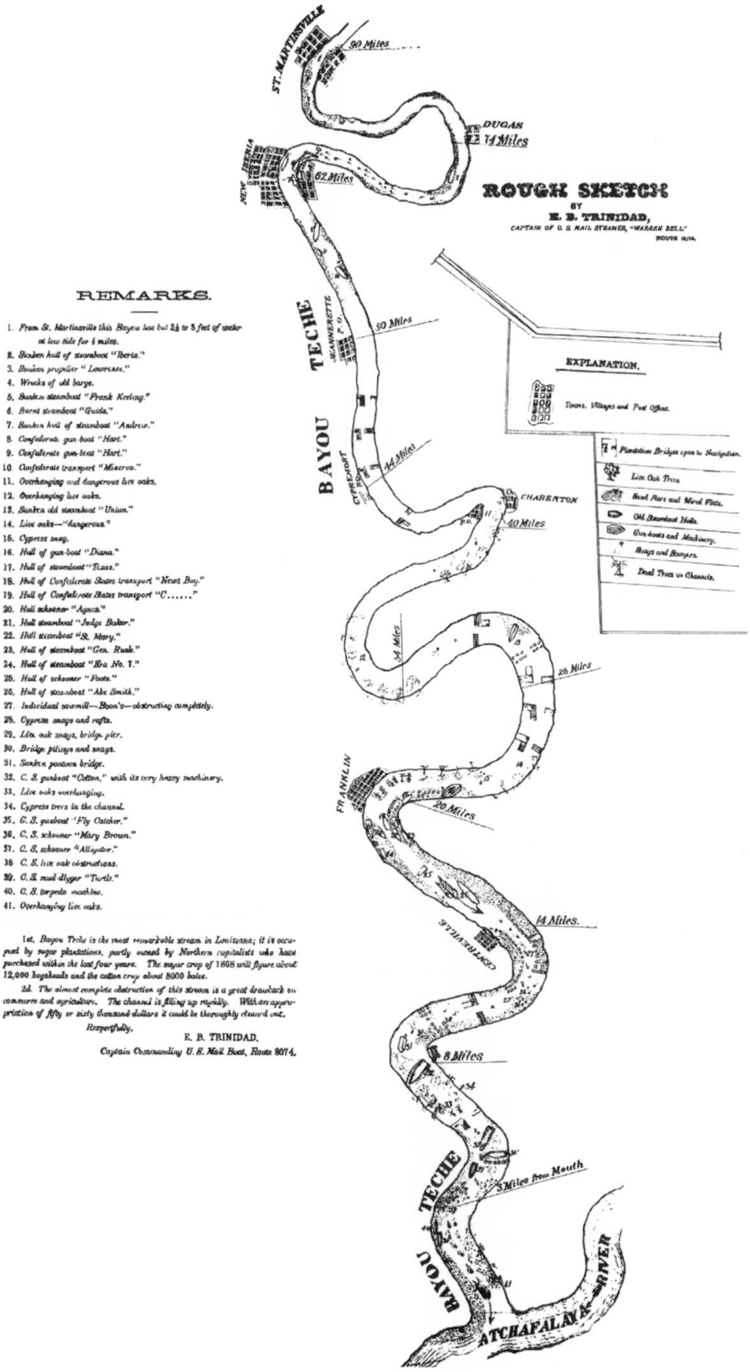

Steamboat captain E. B. Trinidad's "Rough Sketch" map of the Teche, ca. 1868, showing obstructions caused by the Civil War. Source: facsimile, author's collection; original in Cartographic Division, National Archives and Records Administration, Washington, DC.

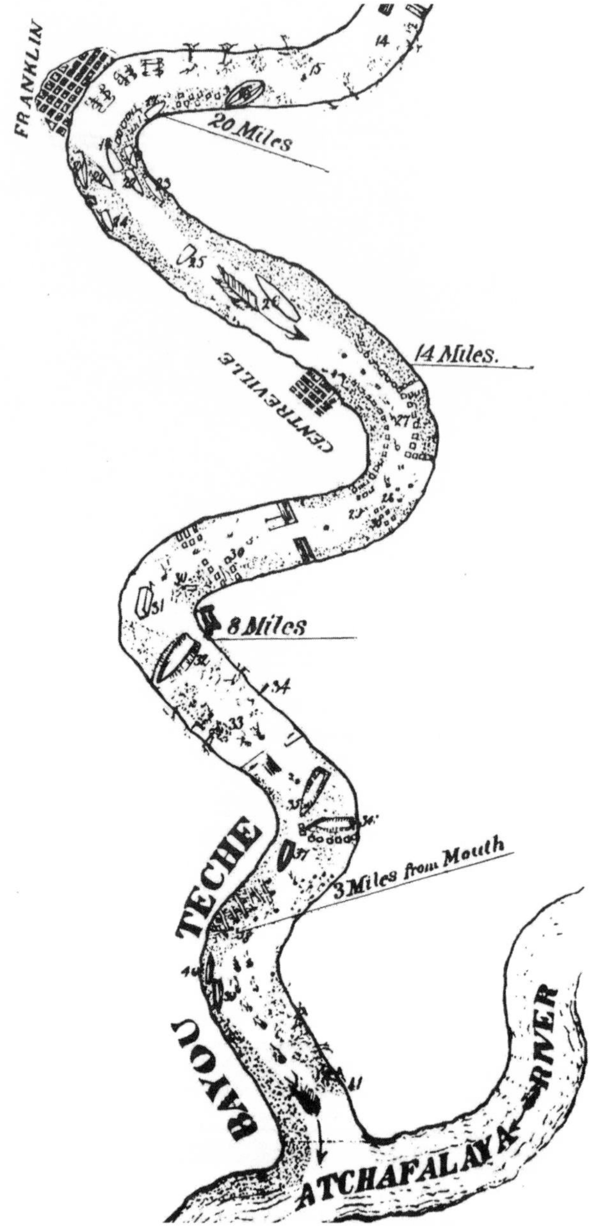

Detail of steamboat captain E. B. Trinidad's "Rough Sketch" map of the Teche, ca. 1868, showing obstructions caused on the lower Teche by the Civil War. Source: facsimile, author's collection; original in Cartographic Division, National Archives and Records Administration, Washington, DC.

Established in 1802, the Corps directed the upkeep of America's navigable waterways. Indeed, no navigable body of water, including the Teche, could be dredged, dammed, widened, or bridged without approval from the Corps and its chief administrator, the U.S. Secretary of War. Although the Corps neglected the Teche prior to the Civil War, it now invoked its mandate to oversee the waterway. Receiving over $17,000 in congressional funding, the Corps sent civil engineer William D. Duke to survey the Teche in May 1870 as a prelude to clearing it. As he paddled a skiff the seventy-five miles from St. Martinsville to the bayou's mouth at Pattersonville, Duke took frequent cross sections and soundings to create a highly detailed thirty-two-sheet map of the Teche. On this map he recorded every stump, log, and sunken vessel that he regarded as a menace to navigation.[2]

"The greatest obstruction" between St. Martinsville and New Iberia, Duke informed the Corps, "is the heavy undergrowth, trees & stumps on both sides, extending in and over the bayou almost intersecting in an unbroken line. . . ." He found the lower reaches of the Teche in similar condition, "very much obstructed by innumerable logs & stumps . . . [and] over-hanging live oaks." There, closer to the bayou's mouth, he also encountered many sunken vessels, including the gunboat *Diana*, "with a portion of her machinery visible—her shaft & flanges, cylinders & old iron." Farther downstream Duke located the remains of the gunboat *Cotton*, which, he noted, "lies nearly at right angles with the stream . . . her stern 6' below low water, [and] her machinery, which is visible, consists of wheel shafts & flanges, cylinders & throttle valves. . . ." Duke's recommendation for handling these wrecks expressed no sentimentality or interest in historical preservation: the vessels, he stated, should be "hauled out or blown up."[3]

On completion of Duke's survey the Corps commissioned another civilian, Daniel M. Kingsbury, to clear the waterway. In February 1871 Kingsbury and his detail of twelve laborers—many of whom, he grumbled, dissipated themselves in drinking and brawling—ascended the waterway in a wrecking flat built especially for the project. Designed much like a snag boat, the wrecking flat boasted a variety of tools for handling unwieldy, possibly submerged obstructions. These tools included grapnel and timber hooks, crowbars, sling chains, ropes, a manual winch, and a twenty-one-foot derrick for lifting heavy loads. Kingsbury christened the flat the *Major Howell*, after his superior, Charles W. Howell of the Corps of Engineers. Regretting this obsequious choice, Kingsbury shortly rechristened the vessel the *Bayou Teche*.[4]

Up and down the waterway Kingsbury and his men removed logs, snags, roots, old pilings, even entire trees growing along the banks—anything that might impede steamboats plying the Teche. The most grueling task, however, would be the removal of sixteen sunken vessels. These included, among others, the *Flycatcher*, *Minerva*, *Gossamer*, *Rob Roy*, *Iberia*, and the gunboats *Diana*, *Cotton*, and *Stevens*. With their hefty timbers, dense machinery, and thick iron plates, the gunboats in particular resisted the wrecking crew's exertions. The ironclads would have to be blasted, just as Duke had recommended earlier, and removed in fragments. "I have made one blast today on the wreck . . . ," Kingsbury reported to Howell regarding the *Cotton*. "The charge was about 75 lbs. powder. It was quite effectual, tearing away a large portion of the front part of the wreck. I shall make one or two more blasts this afternoon."

The *Cotton* proved sturdier than expected and Kingsbury shortly confessed, "I did not think I should have had near the difficulty I have had in destroying her." He reluctantly called in professional divers from New Orleans, who planted waterproof gunpowder charges at weak spots in the wreckage. "We have made 8 blasts since the 9th . . . ," he informed Howell. "One of the blasts[,] the can contained 263 lbs. powder, [and] was placed at the hull directly under her port engine. I think it was very effectual, as large masses of the machinery & timber were thrown high into the air." But, he noted, "[N]early all her timber, as it is blown to pieces, immediately sinks, which will take considerable time for me to remove." Finally, the *Cotton* began to give way. Reclaiming some of her iron, Kingsbury sold it as scrap to fund modifications to the wrecking flat. One intriguing iron artifact may have escaped the melting furnace. "Yesterday we removed large masses of timber from the wreck," Kingsbury recorded. "Among the mass was an iron cannon, which fell off the mass of timber in hoisting the same."

Steamboat crew members and passengers lauded the improvements, even as Kingsbury and his men toiled away on the unfinished job. "The steamboat *Warren Belle* yesterday evening passed up the channel made on the left bank of the bayou going up," Kingsbury told the Corps. "She had quite a large number of passengers on board. They appeared much pleased in seeing the obstructions removed thus far, the ladies and gentlemen waving their handkerchiefs & their hats, accompanied with a pistol salute." By project's end in late 1871, Kingsbury had removed from the Teche, among other detritus, sixteen wrecks (three partially); eighty-two bridge pilings; twenty-one "dangerous snags"; thirty-eight overhanging trees; one hundred six limbs; and a raft of one hundred ninety-one sunken "large live-oak logs." Captain Trinidad gleefully wrote Major Howell, "It is a pleasant duty to acknowledge that the work done, under your instructions, by Captain Kingsbury is complete in every respect, and [has] restored once more that important stream, Bayou Teche, to navigation."[5]

Congress authorized more improvements to the bayou in the River and Harbor Act of March 3, 1879, and in 1880 and 1881 it apportioned $26,000 specifically to improve the Teche north of St. Martinsville. The water level in that stretch often ran extremely low, as the Corps observed in 1880: "[T]here has not been sufficient water for navigation [in the uppermost Teche] within the memory of the oldest inhabitants, except during the extreme high-water of 1874." Another army engineer echoed this account, noting, "The upper 6 miles of the . . . Teche is but little more than a gully; its banks are very steep, heavily timbered, and . . . [d]uring a large portion of each year this part of the bed of the bayou has no water in it except in little pools; in fact this portion of the stream becomes dry."[6]

Still, even the high-water season could leave the bayou dry above St. Martinsville. "[T]here was not enough water in the Upper Teche to permit my examination to be made in a perogue [pirogue]," lamented an army engineer in December 1886, about a month into the high-water season. Moreover, a gauntlet of obstacles would have barred navigation to the head of the bayou. Another army engineer thus described the four miles of the Teche below Port Barre as "filled by standing trees, which grow even down to the bottom of it, and with overhanging trees, bushes, and logs and fallen trees. . . ."[7]

Motivated by appeals from local and state politicians, the Corps resolved in the early 1880s to open the entire upper Teche to commercial traffic. If achieved, the bayou could be used not only by those living on its banks, but by others on the Courtableau and its tributaries who desired a less turbulent path to New Orleans than the wild, sometimes log-choked Atchafalaya River. The Corps would open the upper Teche, it proposed, not so much by deepening its channel through dredging, but by building a series of locks and dams to raise the water level on the bayou. The project, asserted the Corps, would require three sets of locks and dams—one below and two above St. Martinsville. The estimated cost: nearly $90,000, or more than three times the original budget fixed by Congress for Teche improvements. This estimate swelled to over $135,000 once it became clear the project would require dredge boats, derricks, and a small army of laborers for at least a year.[8]

While the federal government mulled over the expense of these proposed locks and dams, the Corps pushed forward with less costly improvements, removing "overhanging trees, logs, snags, and other obstructions" from the thirty-eight miles of bayou between St. Martinsville and Leonville. Four years later the Corps cleared the remaining eleven miles to Port Barre. During an ideal high-water month, estimated an army engineer, a steamboat 175 feet long by 30 feet wide could now ascend the Teche to within ten miles of Port Barre—to within eight miles if not for a static bridge blocking the route.[9]

Ironically, the Corps of Engineers' improvements came just as steamboat traffic on the Teche, as elsewhere, entered a long, slow decline. During the decade and a half prior to the Civil War, steamboats from Bayou Teche arrived at New Orleans an average of 115 times per year. By the early to mid-1880s, however, the average had fallen to a mere 37 arrivals per year. This precipitous drop stemmed from competition with a new mode of transportation, one that threatened to make Bayou Teche obsolete as a conduit for moving people, sugar, cotton, livestock, merchandise, and other things once consigned solely to waterborne shipping. That new means of transportation was the railroad.[10]

In 1857 the New Orleans, Opelousas, and Great Western Railroad connected New Orleans to Brashear City, only about 10.5 miles by water from the mouth of the Teche. Planning to extend this line northwest along the bayou, the railroad company installed a graded rail bed to New Iberia. From there the bed veered northwest from the Teche toward Vermilionville. The Civil War halted the laying of track beyond Brashear City, but construction resumed after the conflict. The railroad—renamed Morgan's Louisiana and Texas Railroad in homage to its owner, shipping magnate Charles Morgan—finally reached New Iberia in 1879. By next summer the railroad connected New Iberia, Jeanerette, Franklin, Centreville, and Pattersonville to such seemingly distant cities as New Orleans and Houston. By 1901 additional rails extended along the Teche as far upstream as Arnaudville.[11]

Predictably, the railroad disrupted the Teche Country's venerable steamboat industry, which at the time consisted of the People's Independent Teche and Atchafalaya Line, Pharr Line Teche Steamers, and the New Orleans and Bayou Teche Packet Company, not to mention any number of itinerant tramp steamers. To the dismay of these steamboat operators, the railroad captured about 90 percent of the bayou's shipping business within six years of its appearance—even while charging about 15 percent more than steamboats per ton of freight. Planters opted for the higher rate, surmised the Corps of Engineers, because the sooner produce reached market, the sooner they received payment. For instance, a train could travel the 87 miles of track between Pattersonville and New Orleans in under five hours. A steamboat, however, took five days to make the same journey via a circuitous, sometimes hazardous route covering over 300 miles—the distance to New Orleans by water since 1866, when Iberville Parish authorities dammed the shorter Bayou Plaquemine route as a flood control measure. (It did not help the Teche steamboat trade when the Pharr Line, hoping to stave off its own demise, colluded with Morgan's railroad to destroy its rival steamer lines.)[12]

Some of the old steamers, however, found new utility moving showboats—which, contrary to their popular image as self-propelled theaters, were actually unpowered barges that relied on smaller pushboats for motion. Showboats had appeared on the Teche as early as the 1840s, when a female promoter named Houston visited the bayou with her "large boat fitted up for the exhibition of wax figures and theatrical performances, and also for the sale of goods. . . ." Showboats plied the Teche more frequently in the late nineteenth century, when these "floating palaces," as they often were billed, ventured from their primary Mississippi River routes to find new audiences. As one showboat historian has observed, the late-nineteenth-century vessels "penetrated deep into the Bayou Teche country . . . a veritable gold coast for the showboat."

Blowing their whistles, striking up their calliopes, and beaming their searchlights at night, showboats ascended the Teche as far as St. Martinsville and sometimes ventured even farther upstream. In the 1840s, for example, Ms. Houston's vessel entertained above St. Martinsville; and in spring 1886 the steamboat *Hattie Bliss* carried a "theatrical troup" [*sic*]—probably "DeVere's Carnival of Novelties and Mastodon Dog Circus," supported by the "Rex Silver Cornet Band"—above St. Martinsville to reach Arnaudville. But wherever showboats moored on the Teche, they attracted throngs of paying locals who relished vaudeville routines and the period's ever-popular melodramas. As a St. Martinsville native recalled, "[T]he ship was primitive, the benches uncomfortable, the red velvet curtain shabby. . . . [But a]s the curtail fell, a happy bedlam broke loose in the audience. They . . . did not want this thrilling entertainment to end."[13]

In the early 1900s the Army Corps of Engineers oversaw two improvements that revived hope for the Teche steamboat trade. First, in 1909 the Corps replaced the Iberville Parish dam on Bayou Plaquemine with a lock, restoring this once popular navigational link between the Atchafalaya and the Mississippi. Significantly, this project reduced the traveling distance between the Teche and New Orleans by 135 miles and made it possible to reach the city from New Iberia in only three days instead of five. Second, in 1913 the Corps finally built a lock and dam on Bayou Teche between New Iberia and St. Martinville—again, to raise the upper Teche's water level for the benefit of steam navigation. Called Keystone Lock and Dam (after adjacent Keystone Plantation), this project traced its origin to the Corps' proposal of over three decades earlier calling for three such structures on the bayou.[14]

Congress had rejected that idea as extravagant, but it eventually approved construction of one lock and dam. A single facility, conceded the Corps, would not raise the Teche sufficiently to permit navigation to Port Barre. "Of these upper 19 miles of the Teche," an engineer reported, "I believe it to be impracticable to improve the 4 or 5 miles immediately below Port Barre for steam navigation, except at an expense greatly in excess of any benefit to be derived from such improvement." In the end, the Teche's uppermost reaches would not be opened to navigation until 1920, when a private business, the Atchafalaya-Teche-Vermilion Company, dredged the upper bayou to Port Barre. The company also excavated Ruth Canal between the Teche and the Vermilion River. (The purpose of these related projects was not to assist navigation, but to irrigate rice farms along those waterways.)[15]

Despite congressional approval, the Keystone project had remained in limbo until passage of the River and Harbor Act of March 2, 1907. In that bill Congress authorized the Corps to secure "a 6-foot navigation to Arnaudville . . . at an estimated cost of $111,000, by dredging, removal of snags, etc., *and construction of a lock*" (emphasis added). To a twelve-acre tract at Keystone Plantation the Corps now brought in steel from Pittsburgh and Boston, gravel and sand from St. Louis and nearby Petite Anse Island (by then renamed Avery Island), and stone, timber, and cement from New Orleans. To claim additional water for the Teche above the dam, the Corps built a smaller dam on Bayou Fuselier and expanded a pre-existing channel, known as Keystone Canal, to divert water from Spanish Lake.

In summer 1913 the Corps completed major construction on the project and opened it for operation—only to find demand for the lock surprisingly lower than expected. Engineers dubiously blamed the lack of interest on World War I and its impact on international shipping. In actuality, the amount of freight passing through Keystone Lock only dropped after the conflict and would not exceed wartime levels until the mid-1930s.[16]

The underutilization of Keystone Lock stemmed from competition, not only from the railroad, but from two new inventions: the automobile (including the versatile motor truck) and the modern highway system. While state and parish governments had created a road along much of the Teche by 1840, it had been designed for foot, horse, and wagon. In the early twentieth century, however, it became clear that motorized traffic would soon become commonplace and that better roads had to be constructed. Underwritten by the Federal Aid Road Act of 1916, the Federal Highway Act of 1921, and private boosters, a modern graveled road reached the Teche Country in the early 1920s. Passing through Patterson, Centerville, Franklin, Baldwin, Jeanerette, and New Iberia, the new highway on its extremes connected the bayou—strange as it may have seemed—to St. Augustine, Florida, and San Diego, California. Officials called the highway U.S. 90, though it was also known as the Southern National Highway and, more frequently, the Old Spanish Trail (all of which overlapped along the Teche with a segment of the Pershing Highway).[17]

By the late 1920s another federal road, Highway 190 (or the Evangeline Highway), crossed the headwaters of the Teche to link Port Barre to the national highway system. In little more than a decade both it and Highway 90 were paved. Another paved highway also ran along the west bank of the Teche from New Iberia through St. Martinville, Parks, and Breaux Bridge to Cecilia. From there a graveled all-weather road continued through Arnaudville and Leonville to connect with Highway 190 between Opelousas and Port Barre. By the eve of World War II, modern roads traced the entire length of Bayou Teche.[18]

Combined with the railroad, the twin threats of the automobile and the highway system severely damaged the steamboat trade. In 1899, right before the age of mass-produced automobiles, the number of steamboats on the Teche stood at twenty—some of which had been reduced to "jobbing" (short temporary work) or hosting passenger excursions to picnics on the seashore. A quarter century later only eleven steamboats remained on the bayou. By then steamboats were rarely used to carry passengers: from the mid-1920s to the early 1930s the annual number of passengers on the Teche averaged only ninety-eight persons. Steamboats repurposed to push showboats suffered a similar decline, as the floating theaters yielded to the railroad circus, the movie theater, and eventually the Great Depression. The few Teche steamboats that remained in use primarily carried the residue left by the railroads: lesser amounts of sugar, molasses, syrup, cotton, rice, fuel oil, and oyster shells (used for surfacing). Still, there was lumber—colossal amounts of lumber.[19]

Although mills along the Teche manufactured this lumber, the raw timber itself came from elsewhere. It had to, because few woodlands remained on the Teche. Many had been cut down for use as firewood in the sugar mills, but substantial amounts of local timber no doubt

2856 – The Bayou Teche, Louisiana,
Sunset Route, Southern Pacific.

Steamboat and log raft on Bayou Teche, ca. 1910. The logs awaited processing at one of the bayou's several lumber mills. Ironically, this postcard was issued by the Southern Pacific Railroad Company, whose success helped bring about the demise of the Teche steamboat trade. Source: author's collection.

also went into construction, not to mention into cast-iron stoves, fireplaces, and steamboat boilers. As early as 1845 a travel writer observed the bayou's "deficiency of ordinary fire-wood." Four decades later F. D. Richardson, who resided along the Teche at Jeanerette, lamented: "Much of the splendid forest scenery has been destroyed, and where once stood the giant live oak . . . [and] the bright, glistening magnolia . . . is now a field, and nothing more." "The timber is practically gone," sighed James P. Kemper of Franklin, who denigrated others along the Teche for "chopping down trees and letting them rot, sometimes just to rob the bees of a few pints of honey. . . ."[20]

Regardless, a lumber industry boom occurred along the Teche in the late nineteenth and early twentieth centuries, and the raw timber came primarily from swamps of bald (red) cypress along nearby Grand Lake. This timber had to be transported from the swamps to the Teche, and it was the old dependable steamboats that did the towing. Lumber mills leased or bought these steamboats, using the vessels to pull lengthy rafts of timber to the bayouside mills. "[O]n this stream are situated . . . some of the largest lumber manufacturing concerns in the State," an observer wrote in 1899, "to whose use the Teche is absolutely indispensable for floating up the timber in rafts."

Although steam-powered lumber mills had operated on the bayou for generations, those applying state-of-the-art, mass-production technology appeared only around 1890. In 1901, for example, a lumber trade journal praised a Jeanerette mill's impressive-sounding "Allis-Chalmers double-cutting telescopic band mill." Only "the second mill of the kind ever run in the South," touted the journal, it annually generated 20,000,000 feet of lumber, 50,000,000 shingles, and 50,000,000 laths (flat, narrow strips of wood used in plaster walls). Added the journal, "The whole plant is equipped [with] the latest known devices for economical handling of the product . . . [and] is strictly modern in every respect. . . ."

By 1909 ten such mills operated on the Teche, including the Iberia Cypress Company and the Gebert Shingle Company in New Iberia; the Jeanerette Lumber and Shingle Company; the Kyle Lumber Company in Franklin; the Albert Hanson Lumber Company in Garden City; not to mention the Riggs Cypress Company and the enormous F. B. Williams Cypress Company, both of which stood near the confluence of the Teche and Lower Atchafalaya at Patterson. Annually, these mills converted about 144 million feet of logs into lumber, shingles, and other wood products valued at over four million dollars.

Through indifference to the environment, however, the Teche lumber industry shortly brought about its own dissolution. What had been dark, lush semitropical forests of towering bald cypress soon became desolate ponds of rotting stumps. As the Louisiana Department of Conservation lamented in 1921 regarding lumber practices statewide, "When the timber ran out, the mill shut down. . . . So far as the local residents were concerned, the lumber industry came, lived an existence of 15 or 20 years, and moved on, with little benefit to anyone."[21]

The lumber industry did not stop or reverse the steamboat's decline on the bayou—but it did help to delay extinction. When the mills finally shut down, steamboats on the Teche nudged once more toward extinction. By the early 1940s only a few steamboats still operated on the bayou. These survivors included the *Albert Hanson*, the *M. E. Norman*, and the *Amy Hewes*. Each continued to tow cypress rafts to the few remaining mills, then, like the mills themselves, vanished. After more than a century of utility, the age of steam had come to an end on the bayou. (The paddlewheeler *V. J. Kurzweg* also plied the Teche into the early 1940s; but she was not a steamboat, relying instead on diesel motors.)[22]

Just as America entered World War II, the Corps of Engineers finished three mammoth projects that drastically changed the character of the Teche. These were the building of the West Atchafalaya Basin Protection Levee; the Charenton Drainage and Navigation Canal; and Wax Lake Outlet. All three sprang from the calamitous Flood of 1927 and the ensuing series of Flood Control Acts passed by Congress to prevent similar disasters. Completed by 1942, the West Atchafalaya Basin Protection Levee ran nearly 129 miles in length: from Avoyelles Parish down to the headwaters of the Teche, then south between the bayou and the Atchafalaya swamp, and finally for several miles past the bayou's mouth at Patterson.[23]

This enormous berm—large enough for its crown to support an unpaved road—forever altered the Teche. It did so by isolating the bayou from its vital fresh-water sources, namely, the Atchafalaya River and, in turn, the Red River and the Mississippi. It also blocked Lake Dauterive and Lake Fausse Pointe (as well as the new Borrow Pit Canal whose excavated soil comprised the levee) from discharging seasonal rainwater into the Atchafalaya Basin.[24]

Corps engineers solved the first problem by installing floodgates in the levee to replenish the Teche from the Atchafalaya. They solved the second by building the Charenton Drainage and Navigation Canal, which diverted water trapped by the levee southwest to the Gulf of Mexico. While earlier irrigation, navigation, and drainage canals, such as Ruth Canal, Joe Daigre Canal, Nelson Canal, and Hanson Canal, had been dug along the Teche, they stretched only fifty to seventy-five feet in width. The Charenton Canal, however, ran about as wide as the Teche itself, and its zigzagging path even usurped a roughly five-mile segment of the bayou before heading off toward open water. The project also forced the Teche to abandon a roughly 1,000-foot section of its historical course and, instead, to jog south and then around a small manmade isle before resuming its original path.[25]

But the post-flood engineering project that made the greatest impact on the Teche was the digging of Wax Lake Outlet, also known as Calumet Cut (after nearby Calumet Plantation). Designed to spare Morgan City from another 1927 flood, the outlet would divert Atchafalaya floodwaters away from the city and into the Gulf. It was a monumental task: stretching over thirteen miles in length, the finished outlet measured 45 feet in depth, extended up to 200 feet in width at its bottom, and spanned about 630 feet in width on the water's surface. It demanded the excavation of over 22 million cubic yards of earth and required heavy railroad and highway bridging. Significantly, the outlet also cut across the course of the Teche, which it dammed with massive east and west floodgated levees. This design forced the Teche to abandon about a mile of its historical course. (Unfortunately, the outlet plowed right through a portion of the Fort Bisland battlefield, apparently destroying whatever remained of the earthen fort, a segment of entrenchments, and the probable resting site of artifacts from the gunboat *Cotton*—illustrating to posterity the value of environmental impact studies.)[26]

The West Atchafalaya Basin Protection Levee, the Charenton Drainage and Navigation Canal, and Wax Lake Outlet—these Corps of Engineers projects are largely responsible for the Teche as it exists today. Add to them the Corps' building of Keystone Lock and Dam, and the Atchafalaya-Teche-Vermilion Company's dredging of the upper Teche, and the story of the bayou for the past century is clearly one of designing the waterway for human convenience. In the process engineers isolated the Teche from its natural freshwater sources, bisected it with manmade outlets and levees, and severed it from its historical mouth. The tradeoff, however, was a more docile bayou—one that promoted commerce and resisted major flooding.

And despite the end of the steamboat era, commerce continued to thrive on the Teche into the mid- to late twentieth century. Development of small, powerful, and fuel-efficient diesel engines, light but durable welded-steel hulls, and new propeller and rudder systems revitalized waterborne shipping. In addition, the Intracoastal Waterway, completed along the Gulf Coast in 1949, connected 3,000 miles of inland water routes—including, indirectly through canals and the Lower Atchafalaya River, the Teche. These advances cumulatively gave rise to the modern pushboat-and-barge system, which competed ably with railroads and highways. Significantly, the small "canal-class" pushboats used on the Teche required only four to seven crewmembers, even while moving a fully loaded barge weighing 1,500 tons (compared to the average 990-ton weight of a fully loaded steamboat, which required about twenty-five crewmembers). Boosted by these improvements, shippers saw their freight totals on Bayou Teche rise from an average of 229,200 tons per year in the late steamboat period (1920–35) to 950,630 tons near the end of the century (1980–95). Sugar and molasses remained major sources of that tonnage. Yet barges also carried sizable amounts of crude oil, gasoline, limestone, sand, gravel, metal pipe and tubing, iron ore, marine shell, and other commodities often associated with south Louisiana's oil industry.[27]

These pushboats and barges, however, ultimately shared the Teche with other vessels, including quarter barges (used by marine laborers as living quarters), spud barges (floating platforms used for marine construction), and commercial fishing vessels, most notably shrimp boats. Prompted by the search for oil in nearby wetlands, floatplanes also appeared on the Teche, taking off and landing, for example, at the New Iberia facility of Pelican Aviation Corporation—described in the 1970s as "a ramp sloping into muddy water, a dock and a couple of old hangers shaded by overhanging branches against the hot southern sun." Pleasure craft, however, increasingly outnumbered these more utilitarian forms of transportation. Pirogues (increasingly made of aluminum, fiberglass or plastic instead of traditional cypress), canoes, kayaks, jon boats, bass boats, houseboats, pontoon boats (also known as "party barges"): these became the most common types of vessels on the Teche by the late twentieth century. And with these pleasure boats came water-skiers, tubers, wakeboarders, and personal water craft riders.[28]

It would be easy to say that so too came litter, pollution, and other issues harmful to the present-day Teche. But these problems had already afflicted the Teche for generations, even as far back as the colonial era. (The frontier indigo industry, it will be remembered, produced an extremely toxic byproduct that colonists had to dump somewhere.) Recently, however, government agencies and grassroots activists came together to overcome these challenges—with such devotion that, as will be seen, the Teche in the early twenty-first century seems poised to undergo a renaissance.

Conclusion

After roughly two millennia of Native American occupation, Bayou Teche enticed Europeans to explore its murky waters around the mid-eighteenth century. Behind early French pioneers came the Spanish, followed by Anglo and Scots-Irish settlers along the bayou's lowest stretches. With them were enslaved Africans and the *gens de couleur libre* (free persons of color). For a century and a half the meandering Teche served these explorers, settlers, and slaves, as well as subsequent travelers and entrepreneurs, as a primitive superhighway. It guided them from the soggy littoral to the heart of one of the most fertile, most hauntingly beautiful, sections of the colony, territory, and state.

The region's first major industry, cattle ranching, flourished; the second, indigo planting, failed. Cotton production, however, took to the upper Teche and, eventually, sugar growing, to the lower Teche. Sugar, slavery, and the invention of steam power coalesced on the bayou around 1830. These factors transformed the Teche from a backwater into a thriving "sugar bowl" rivaled only by bayous Lafourche and Terrebonne, and exceeded only by the lower Mississippi River between Baton Rouge and New Orleans. The region's enormous wealth in part attracted the marauding armies and dueling ironclads of the Civil War, devastating the bayou's banks and making its course unnavigable. This destruction left the Teche Country less resilient to the floods, epidemics, and bankruptcies of the postwar era. It also contributed to the sense of despair and fear that vented itself in the period's often racially motivated violence.

Cotton and sugar planting endured along the bayou, though joined by forays into rice and lumber. The advent of the railroad and eventually the motor vehicle and highway stimulated the region's economy—even as these innovations doomed the steamboat and, seemingly, the bayou to insignificance. Predictably, the steamboat did become extinct, both on the Teche and elsewhere; but the same ingenuity that created the railroad and highway soon reinvented river transportation, producing the new pushboat-and-barge system on rivers across America. On the Teche this new system ultimately generated a net increase in tonnage of goods conveyed up and down the bayou; yet it required only a fraction of the boats and deckhands that once moved cargo on the waterway.

With fewer boats and hands—and fewer whistles and searchlights coming 'round the bend—the Teche clearly seemed to lack the vitality of earlier times. Indeed, by the mid-twentieth century the bayou struck some as downright moribund. "[T]he Teche is the Past in Louisiana," author Harnett T. Kane commented sympathetically in the early 1940s. "It has seen and heard excitements. . . . Now it has settled down to a period of serene rest." It was a serenity imposed partly by the construction of so many levees, floodgates, locks, and other water-control structures. These enormous projects changed the Teche from a force of nature into a more obedient stream, one incapable of disrupting everyday life and commerce with the occasional raging deluge.

Shaped by these levees, floodgates, and locks, the story of the Teche in the twentieth century was one of designing the bayou to human advantage. Its story so far this century has been one of conservation in the face of mounting environmental threats, including litter, pollution, overdevelopment, and the spread of invasive species. Local activists have responded to these challenges by forming three groups, namely, the TECHE Project, Cajuns for Bayou Teche, and the Tour du Teche—all of which coalesced around 2009. Of these, the TECHE Project has predominated: since inception it has worked to make the Teche "a healthier waterway for fishing, kayaking, canoeing, boating, tubing and even swimming." It has done so, for example, by cleaning the Teche, and by teaming with the Sierra Club's Delta Chapter to monitor water quality through the Bayou Teche Water Sentinels program. The TECHE Project has also hosted "Trash Bash & Boogie" events, offering complimentary food and music to volunteers who clean the bayou. And it has taught Teche-area landowners to construct nesting boxes for wildfowl living on the waterway and to use native trees and plants to stop bayouside erosion.[1]

Moreover, the TECHE Project has teamed with the National Park Service, the Army Corps of Engineers, the Sovereign Nation of the Chitimacha, and the University of Louisiana at Lafayette's Center for Louisiana Studies, among other entities, to create the Bayou Teche Paddle Trail. The trail's purpose is to instill "a greater appreciation for the ecological, cultural and historic preservation in the [Bayou Teche] watershed." Among other improvements, this program will renovate existing boat launches and docks along the bayou, install a series of new floating docks, and erect directional signs and informational kiosks. In early 2015 the U.S. Department of the Interior chose the paddle trail as one of only eighteen in the National Water Trails System, a network of water trails chosen because of their natural and historic importance. "We worked with community leaders to submit the bayou for this recognition," explained a TECHE Project volunteer, "because it is a great economic driver for tourism, [and] it's another way to celebrate the culture and the nature of South Louisiana."[2]

The TECHE Project also fostered Cajuns for Bayou Teche, which serves as an environmental "brown-water navy," rallying flotillas of private boats to scour the Teche for litter. "I see a lot of dump sites [along the Teche] specifically consisting of household garbage," explained the founder of Cajuns for Bayou Teche. "Residents simply walk to Bayou Teche and throw or burn their trash at the banks. This garbage always ends up in the bayou and in the back yard[s] of other residents." As a program of the TECHE Project, Cajuns for Bayou Teche removed nearly forty tons of garbage from the waterway, including tires, oil barrels, hot water heaters, and refrigerators. It incorporated as an independent organization in 2014.[3]

Similarly, the TECHE Project nurtured the Tour du Teche "adventure race," which incorporated as a distinct non-profit in 2012. The race promotes the bayou "to paddlers and other eco-tourists" while demonstrating to locals "the recreational, aesthetic, cultural and economic value of Bayou Teche." Beginning as a single nonstop race from Port Barre to Berwick (one crew finished in about 18.5 hours; others took over two days), the Tour soon evolved into a three-stage race held in the fall and a series of shorter races held in the spring. The group even hosts the Petit Tour du Teche, a series of races for children. "We're teaching water safety, paddling technique, and environmentalism," explained a parent-supervisor with the Petit Tour du Teche, "but we're all having fun doing it."[4]

During the same period others made important advances in understanding the history of Bayou Teche, particularly its Native American heritage. Explorers knew for some time that ancient Native American sites stood along the Teche. As early as 1784, for example, geographer Thomas Hutchins noted of the bayou: "[I]t is 10 leagues from its mouth to an old Indian village, on the east side, called Mingo Luoac. . . ." In 1913 self-taught archaeologist C. B. Moore mounted the first systematic inquiry into Teche archaeology, traveling up the bayou aboard the steamer *Gopher* to examine prehistoric sites, including two ancient mounds on the bank in Loreauville.[5]

Other archaeologists followed, including Jon Gibson of the University of Southwestern Louisiana (now the University of Louisiana at Lafayette), who in the 1970s conducted an important survey of Native American sites along the Teche Ridge and in the nearby Atchafalaya Basin. In 2003 Mark A. Rees of the University of Louisiana at Lafayette launched the Plaquemine Mounds Archaeological Project, "designed as a means of redressing the need for additional study of mound sites in the western Atchafalaya Basin"—including along the Teche Ridge. (*Plaquemine* in this context refers not to the town of Plaquemine or to Plaquemines Parish, Louisiana, but to a particular Native American culture that thrived in the Lower Mississippi Valley as early as about 800 years ago.) During the course of his project, Rees catalogued seventy-four mound sites in the four Teche parishes. Some of these mounds, such as the purported village sites of *Hi'pinimtc na'mu* and *Okû'nkîskîn* (or *Ama'tpan na'mu*), sat near but not on the Teche. Others, including the purported village site of *Qiteet Kuti'ngi na'mu* and the Loreauville mounds, sat directly on the bayou. Indeed, part of *Qiteet Kuti'ngi na'mu* is slowly eroding into the Teche.[6]

Also in 2003, Rees headed a two-week archaeological field school on the bayou near Loreauville, excavating the dwelling site of Acadian pioneer Armand Broussard. From this sprang a larger endeavor, also involving Rees, called the New Acadia Project. Beginning in 2013 Rees sought the home sites and graves of Acadian exiles who settled along the Teche in 1765—including those of Acadian guerrilla leader Joseph Broussard dit Beausoleil—as well as any artifacts those Acadians left along the bayou. Commercial and residential construction lent urgency to the investigation. "We're losing the archaeological information," Rees explained. "We're losing our heritage. . . . [because] some sites are being destroyed. . . ."[7]

Rees was not the only modern archaeologist to investigate the Teche. From the 1980s through the early 2000s, commercial archeology firm R. Christopher Goodwin & Associates received several contracts from the Army Corps of Engineers and the state of Louisiana to examine the Teche Ridge—and even the bottom of the Teche itself. Goodwin, for example, evaluated the site of Keystone Lock and Dam; a roughly twelve-mile stretch of the Teche between Belle Place and southeast New Iberia; a roughly thirty-mile stretch of the Teche between Jeanerette and Franklin; the Civil War battlefield near the former site of Fort Bisland; and, to quote the title of a Goodwin report, "Magnetic Anomalies Located in Lower Bayou Teche, St. Mary Parish, Louisiana."[8]

Other present-day archaeologists have also conducted research along the Teche. Their projects included examinations of Shadows-on-the-Teche plantation home, Lutzenberger Foundry, and a sunken nineteenth-century steamboat in the Teche, all in New Iberia. (Some have tentatively identified this vessel as the Confederate cottonclad *Teche*, later captured and converted into the Union tinclad *Tensas*.) More recently, archaeologists have proposed a study of Promised Land, a purported slave cemetery on the bayou at Parks. Most of these archaeological projects, however, consisted of extensive surveys involving little if any excavation—meaning much remains to be discovered along the Teche. Looking at underwater historical archaeology alone, for example, researchers have identified 137 shipwrecks in Bayou Teche. Of these, only the purported wreck of the *Teche/Tensas* has been examined in detail.[9]

Meanwhile, other entities worked to promote and conserve the waterway. In 2001 the U.S. Department of the Interior created the Bayou Teche National Wildlife Refuge, consisting of six noncontiguous areas ("units") on both sides of the Teche from Franklin to Centerville. In 2006 Congress established the Atchafalaya National Heritage Area, encompassing the entire course of Bayou Teche, to "increase public awareness of and appreciation for the natural, scenic, cultural, historic and recreational resources of the area," among other benefits. Local governments and organizations also became involved in preserving and promoting the Teche. In 2004 Franklin hosted its first Bayou Teche Black Bear Festival, which included guided boat tours of the waterway and the nearby National Wildlife Refuge; and in 2010 it held the first Bayou Teche Wooden Boat Show, featuring a parade of often handcrafted wooden vessels. During the same period New Iberia established the Bayou Teche Museum to collect, display, and interpret Teche Country artifacts, while Breaux Bridge founded the Teche Fest to raise funds for its Teche Center for the Arts. Even the Catholic Church became involved, organizing a Eucharistic procession on the Teche, called the *Fête-Dieu-du-Teche*, that started at Leonville and visited Arnaudville, Cecilia, Breaux Bridge, Parks, and St. Martinville. "Traveling down the Teche is highly symbolic, because it replicates what our ancestors did," noted the attending bishop. (Pope Francis granted a plenary indulgence to those who participated in the procession.)[10]

Given this burst of recent activity, it seems clear the Teche is poised to undergo, if not actually undergoing, a renaissance. Yet serious challenges remain to the Teche's welfare, including the proliferation of invasive species, both in the Teche and along its banks. One of these species is an infamous nuisance animal: the voracious, burrowing rodent known as the nutria. Other invasive animal species that might soon infiltrate the Teche, if they have not done so already, are the zebra mussel and the Asian carp. Invasive botanical species along the Teche include the Chinese tallow, elephant ear, hydrilla, and giant salvinia—all originally introduced as ornamentals. While some of these botanical species are innocuous when grown in backyards, they become pests in the wild because they rapidly displace native species.[11]

The most vilified of all now-feral ornamentals, however, is the water hyacinth. It has plagued the Teche and other American waterways for over a century, growing into enormous tangled rafts that block navigation, ensnare propellers, and deprive water of oxygen essential to native plants and animals. According to one oft-cited tradition, water hyacinths came to the United States during the Cotton Centennial

That detested invasive species, the water hyacinth, seen here along Irish Bend. Source: photograph by the author.

Exposition of 1884–85, when Japanese exhibitors handed out the plant to admiring visitors. ("Why the Japanese selected a South American plant to give away has been lost to time," a modern journalist aptly observed.) Regardless of its origin, the water hyacinth spread to Louisiana's waterways by 1899, when Congress allocated $25,000 "for the construction of a boat suitable for operating on the navigable streams of Louisiana and removing the aquatic plant" by "raising the plants [out of the water] and crushing them." When this did not work, the federal government rigged vessels to spray the hyacinths with a noxious mixture of arsenic and sal soda (the latter compound helping to disperse the toxic arsenic—which, as a chemical element like carbon or hydrogen, would never break down over time into safer components).[12]

This poisonous spray failed, too, for while the mixture appeared to kill the hyacinths, the plants inevitably returned in larger quantities than before. (The spray did cause "the death of a large number of cattle," St. Martinville's newspaper reported in 1914.) Another eradication idea, proposed in 1912 by Congressman Robert Broussard of New Iberia, never came to fruition: he advised the elimination of water hyacinths

Trash in the Teche, 2012. Source: photograph by the author.

through the introduction of hippopotamuses to south Louisiana's bayous. "[T]he water hyacinth is a favorite food of the big water beasts," a newspaper explained of Broussard's scheme. Moreover, surplus hippos could be hunted, butchered, and eaten like any other wild game. Broussard seemed unaware, however, that hippopotamuses are extremely aggressive and rank among the world's most dangerous animals. (As a humorous nod to Broussard's idea, the TECHE Project refers to its water hyacinth control program as "Project Hippo." The group, however, rounds up the invasive plants manually and mechanically.)[13]

Another perennial problem for the Teche is litter. As a journalist observed in 2009: "Beautiful as Bayou Teche is, under its serene surface lies a monster, the mother of prop killers and fly line snarlers. What could be so powerful it could stop a cigarette boat in full throttle? Neither Grendel nor his mother, not the kraken or Charybdis. What lurks on the bottom of the waterway is garbage. Untreaded tires, cast-off hot water heaters, crab traps, coke bottles. . . ." As noted, groups like the TECHE Project and Cajuns for Bayou Teche have targeted litter and in doing so have significantly beautified the waterway. Yet litter has an insidious way of recurring. Furthermore, much of the litter defiling the

Teche Country is found not on the bayou itself, or on its banks, but on the country roads that border the waterway. The state of Louisiana has designated this series of roads the Bayou Teche Scenic Byway. It is scenic, but, as local novelist James Lee Burke has observed, splendor often lies next to rural blight:

> At sunset, Bayou Teche is high and dark from the spring rains; the air smells of gardenia and magnolia; and antebellum homes glow among the trees with a soft electrical whiteness. . . . But inside that perfect bucolic moment, there is another reality at work, one that doesn't stand examination in the harsh light of day. The rain ditches along that same road are strewn with bottles, beer cans, and raw garbage. Under the bayou's rain-dented surface lie discarded paint and motor oil-cans, containers of industrial solvents, rubber tires, and construction debris that will never biologically degrade.[14]

The most serious present-day threat to the Teche is pollution. It is not, however, a new concern. Teche Country residents expressed anxiety about unclean water over a century ago. Between 1910 and 1911, for example, a blue-ribbon panel of physicians, engineers, and businessmen steamed up and down the bayou to inspect sanitary conditions and collect water samples. One of the panel's findings: the bayou's fish and plants were being destroyed by muriatic acid and caustic potash, toxins used in industrial sugar refining. This contamination, the panel reported, gave rise to "the most disgusting stench" from "rotting fish and vegetable life[,] . . . charging the air to a great distance about with an odor so offensive as to be sickening." Sugar refineries, however, did not bear all the blame. The panel found residents up and down the Teche sending raw sewage from humans and livestock directly into the waterway. "The Teche for all practical purposes," remarked the panel, "is used as a cheap and convenient sewer. . . ." Yet farm animals still drank from the bayou, as did sometimes the poor living on the banks—explaining, observed the panel, the high number of Teche Country typhoid cases (between 500 and 700 annually).[15]

Despite improvements, the Teche's health remained precarious into the late twentieth century. In 1975, for instance, the *Times-Picayune* reported that state wildlife officials had investigated "a fish kill covering

20 miles of Bayou Teche . . . and blamed it on waste flowing from sugar mills." Another fish kill occurred in 1982, when an eyewitness told the media, "The water turned from its usual chocolate brown to black, and dead fish were floating belly-up all over." Officials again blamed the sugar refineries, flatly declaring, "What they're doing is killing the bayou." But agricultural runoff, some making its way into the bayou from distant farmlands, was an additional culprit. As late as 1999–2001, the U.S. Geological Survey analyzed a fish sample from Bayou Teche that contained levels of the pesticide DDT 75 percent higher than the threshold considered unsafe. (Although the federal government banned DDT in 1972, the toxin has persisted in the foodchain.)[16]

Water-quality tests from the same period also found unusually high levels of DDE (formed by the breakdown of DDT); carbofuran (another pesticide, banned in 2009 for use on domestic food crops because of its extreme toxicity); the elements arsenic—possibly left over from fighting water hyacinths a century ago—chromium, copper, and zinc (all poisonous in high concentrations); PAHs (toxic compounds formed by the incomplete burning of organic compounds like coal, oil, gasoline, and even garbage); and pathogens (namely fecal coliforms—that is, bacteria like *E. coli* found in human and animal feces) released into the bayou by faulty rural septic systems, among other sources. All these substances can harm living creatures, causing cancer, nerve damage, and genetic mutations, among other diseases. As a result, between 1998 and 2002 the U.S. Environmental Protection Agency (EPA) declared nearly the entire Teche an "impaired waterway," meaning "at times, it is unfit for fish survival and recreational activities." As late as 2012—the most recent year it assessed the bayou's health—the EPA listed the Teche from Port Barre to Wax Lake Outlet as "impaired" for fish survival and water recreation, including swimming and even boating. (This explains, for example, public signs on the bayou at New Iberia and Jeanerette warning "Water Quality May Not Be Suitable for Swimming.") The Teche, however, is not under a fishing advisory, and the public may boat or swim freely in the bayou, albeit at its own risk.[17]

Despite these shortcomings, the outlook for Bayou Teche is increasingly positive. In recent years the governments of all four Teche parishes joined a Louisiana Department of Environmental Quality (LDEQ)

Water-quality warning sign, Jeanerette, 2012. Source: photograph by the author.

program to reduce "nonpoint source pollution" in the bayou—that is, pollution that originated outside the Teche Country, possibly even outside Louisiana. In fact, St. Landry and St. Martin parishes both enacted a Municipal Separate Storm Sewer System program to reduce discharges of polluted water into the Teche. Moreover, the LDEQ and its Teche-area watershed coordinator teamed with parish leaders, mayors, civic groups, and grassroots organizations like the TECHE Project to draw up a Watershed Implementation Plan. The plan's goal is to restore the bayou's health, expunge it from the federal list of impaired waterways, and keep it off the list by boosting public conservation through educational outreach. To advance this effort, Whitney Broussard of the University of Louisiana at Lafayette's Institute for Coastal Ecology and Engineering monitored water quality at fifteen points along the Teche for an entire year, identifying problem hot spots where, for example, the amount of dissolved oxygen registered troublingly low or the amount of fecal coliforms seemed abnormally high.[18]

Further restoration of the Teche will require government and grassroots activists to overcome any number of obstacles, from lack of cooperation between the upper and lower reaches of the Teche—Port Barre, Leonville, and Arnaudville, for example, sometimes seem worlds apart from Franklin, Centerville, and Patterson—to financial and bureaucratic issues, to pushback from those who do not share their visions for the bayou. When, for instance, the TECHE Project suggested the state legislature declare the Teche a Louisiana Historic and Scenic River, many landowners along the waterway objected because of property rights concerns. Despite their differences, both factions agreed on the Teche's importance as a natural resource. As an opposition leader noted, "I've lived on the banks of that bayou for 59 years and nobody loves it more than I do." He added, "We are willing to assist with clean-up efforts in any way we can." And those clean-up efforts are working. "It is coming back," the executive director of the TECHE Project observed in 2014. "The water quality is good again. It has come a long way." As an ecologist, Broussard concurred, stating, "We are moving in the right direction, but . . . we cannot let up on the progress that has been made. . . ." He added, "Bottom line is that the Teche is much better than it was . . . even 10–20 years ago. . . . I expect that we are 10–20 years away from a healthy bayou similar to its glory days before the industrial revolution."[19]

Bayou Teche belongs to everyone: this is not mere sentiment but state law, which holds that "all navigable waters and the beds of same within its boundaries are common or public things and insusceptible of private ownership." At the same time, much of the land along the Teche is privately owned. If "the Teche" is regarded not merely as the bayou, but as the bayou and the landscape through which it runs—the Teche Country, as this book has called it—then it stands to reason that the Teche's welfare depends on a public-private alliance. This alliance will undoubtedly spark tension, not only between public and private sectors, but between developers and conservationists. Indeed, these tensions are already evident: traveling the bayou today reveals venerable live oaks, stands of mature bald cypress, and preening wildfowl beside industrial sites, convenience stores, and modern suburban housing. Yet lengthy segments of the Teche remain untouched or at least bucolic—the windswept fields of ripening sugarcane, the shaded bayouside cemeteries of whitewashed shrines, the stately antebellum homes that have survived war, floods, and neglect. It is time to decide how much of the Teche should be spared from the advance of modern sprawl, else, like the proud, ornate steamboats that once commanded the bayou, the beauty of the Teche might exist only in memory.

Part Two

Teche Canoe Trip Journal (2011–13)

To research my history of Bayou Teche, I resolved to paddle the entire length of the waterway. Friends found this humorous because they knew me to be an "avid indoorsman." But how could I not paddle it? Had I not paddled it, someone, somewhere—at a book signing, during an interview—would inevitably ask me, "Have you ever been on Bayou Teche?" to which I would have had to answer, "No—but I drive across it every day on the way to work."

So I decided to explore the entire bayou and to do so not by motorboat (which seemed like cheating), but the old-fashioned way, by canoe—all approximately one hundred twenty-five miles of it. It seemed the thorough thing to do, if I were going to write a book about the Teche. And this proved correct, because there was no substitute for seeing for oneself, from a canoe, where the Teche springs from Bayou Courtableau, meets Bayou Fuselier, zigzags at Baldwin, juts out at Irish Bend, or pours into the Lower Atchafalaya.

I would not canoe the bayou all at once, however, but in stages over many months. And I would do so slowly, stopping to take ample notes, photographs, and GPS coordinates. This journal is based on my field notes from that trek.

I drafted two friends to assist me, both recent archaeology graduates from the University of Louisiana at Lafayette: Preston Guidry and Jacques Doucet. Preston's father, Keith Guidry, made up a last-minute addition to the team. And while Preston and Jacques ended up sitting out several stages of the trip, Keith became my one constant. We, the two older guys in our forties and fifties, would be the only two members of the canoe team to complete the entire journey. Ben Guidry filled in for his brother Preston on a couple stages; and my

Our journey down the Teche began at this Bayou Courtableau boat ramp, October 2011. Back, left to right, Jacques Doucet, Preston Guidry, Shane K. Bernard; front, Keith Guidry. Not shown: Ben Guidry, Donald Arceneaux. Source: photograph by the author.

fellow history enthusiast Donald Arceneaux joined Keith and me for one stage.

I was the least experienced paddler of the group, but the others made up for my greenness. Together, an Arceneaux, a Doucet, three Guidrys, and a Bernard, we would seek an answer to the weighty question, "How many Cajuns does it take to paddle Bayou Teche?"

Stage 1: Port Barre to Arnaudville

We began our trip around 9 a.m. on October 23, 2011. The temperature that morning was cool (about 56°F), but grew warmer in the afternoon (about 81°F). The sky was mostly clear. We chose that day to start our trip because it fell shortly after the annual Tour du Teche canoe race. We didn't want to get in the way of the racers and so let them go first.

Preston, Jacques, Keith, and I put in at the public boat ramp in Port Barre [30.558367, -91.954874],* on Bayou Courtableau. Port Barre—the town whose speed trap the swamp pop band Rufus Jagneaux sang about in the early 1970s:

All of y'all know about Port Barre.
(If) they catch you there it's half of your hide.
You might as well gone to Tucumcari.
It costs you that to leave there alive.

In all fairness to Port Barre, the song was actually written about a speed trap in nearby Krotz Springs; but "Port Barre" sounded more lyrical to the composer.[1]

It was in Port Barre that the Courtableau Inn, a nightclub owned by my great-grandfather, Oscar Bordelon—first name pronounced OH-SCARE in the French manner—stood right across the bayou from the present-day boat ramp. In fact, it stood on the lot now occupied by the Squeaky Clean Car/Boat Wash. It was there, in that nightclub in the late forties and early fifties, that my father heard Cajun musicians like Papa Cairo and Nathan Abshire. Dad recalls slot machines lining the walls, even though the devices were illegal at the time. But those were the days of legendary St. Landry Parish sheriff "Cat" Doucet, who tolerated gambling and certain other prohibited vices.

Within minutes of embarking we were paddling on the Teche itself. At Port Barre the bayou is narrow, varying from about 75 to 95 feet wide (not much larger than a sizable *coulée*, our regional word for a ravine). Dense tree canopies reached toward the opposing banks and pressed in on our canoes. Before it was dredged around 1920 this stretch of the

---

* All geographic coordinates are rendered in the decimal degrees format and are compatible with the Internet site Google Maps (www.google.com/maps).

The nonprofit TECHE Project sells a detailed, approximately 30x12-inch printed map of Bayou Teche on waterproof paper for $20.00 + postage. One side contains a navigation map, and the other a historical and cultural map annotated by the author. For more information, visit www.techeproject.org/bayou-teche-paddle-trail/map/ or contact techeproject@gmail.com.

Keith Guidry and Shane K. Bernard (left to right) paddling the upper Teche. Source: photograph by Jacques Doucet.

Teche was navigable only during high-water season (each December to June) or if a sizable flood had occurred recently.

The current at Port Barre was swift and strong that day, even though bayous are known for lethargic, even nonexistent currents. With casual paddling it carried our two canoes downstream at a decent clip. This fooled us into thinking we would make similarly fast headway the entire length of the bayou; but farther downstream the current slowed and, eventually, reversed, reducing our progress until paddling became a more laborious task. By the fourth or fifth stage it took us ten or twelve hours to accomplish what we had originally achieved in half that time. But this lesson would not be learned for several months, finishing, as we did, a stage every month to month and a half.

Shortly after leaving Port Barre we spotted a coyote running along the east bank—our first sight of wildlife. We soon saw a number of large bleached bones on the west bank, probably from cattle. Right before that we caught a strong scent of cattle dung and must have passed a ranch (or *vacherie*, as our ancestors would have called it). We would pass other ranches during the several stages of our trip. On one occasion Keith shouted an insult at a bull staring insolently at us. The

Examining the remains of a vintage automobile on the upper Teche. Note the high-water mark on the vehicle. Source: photograph by the author.

bull took a step toward us and stomped a hoof. That much is true, but the story became more elaborate with each retelling, and after several months Keith had the bull jumping in the bayou to pursue us.

There would be plenty of joking on the trip, including recurring riffs on *Deliverance* (explanation unnecessary).

One time that day Keith and I stopped our canoe beside a bridge to wait for Preston and Jacques. As we floated there, someone threw a bottle from a passing car. Clipping the rail, the bottle shattered in its anonymous paper bag and splashed in the water beside me. How much trash has ended up in the Teche in that manner?

Not too long after seeing the coyote, Preston and Jacques spotted a water moccasin swimming in the bayou. Surprisingly, it was the only snake we saw during the many stages of our trip. Likewise, we saw only one alligator, and it was dead. Actually, make that two alligators, because we spotted a large one on our final leg of the journey. I'll describe that incident later.

I noticed much less garbage on the bayou than expected, certainly a tribute to the efforts of the TECHE Project, Cajuns for Bayou Teche, and the Tour du Teche. The largest quantity of garbage I saw that day lay inside the city limits of Port Barre, where someone used the banks of the bayou as a private garbage dump. Outside of town, however, the garbage became less frequent. Some of it was vintage, which gave it a certain respectability: no one wants to see a late-model clothes washer or automobile half-submerged in the bayou or protruding from its banks; but a rusty Depression-era clothes washer or vintage 1950s coupe are more interesting, and even appealing in their antiquity.

Paddling down the Teche that day we passed the mouths of a few Teche tributaries—so minor that my canoe team and I missed them entirely. Or if we saw them, we must have disregarded them as mere ditches. In any event, the first of these small waterways was Bayou Toulouse, which flows into the Teche where the Union Pacific (formerly Missouri Pacific) railroad spans the bayou just below Port Barre. The next tributary was Bayou Little Teche, formerly known as Bayou Marie Croquant. Then came Bayou del Puent[e].

In this region we passed the cluster of rural houses known as Notleyville. Like this place, many other unincorporated communities—named yet almost imperceptible to outsiders—stand along the length of the Teche, from Bushville in St. Martin Parish to Hubertville in Iberia Parish, to name only a couple. In fact, just southwest of Notleyville we reached the fertile expanse of farmland known to residents as Prairie Laurent. Nearby lay other rural districts known even today to locals by their traditional French names. A short distance downstream from Prairie Laurent, for example, sits Prairie des Femmes (Prairie of Women—the name loses its poetry in translation); then comes Prairie Basse (Low Prairie); and, on the opposite bank of the Teche, the verdant plain known as Prairie Gros Chevreuil (Big Deer Prairie). These names go far back in local history. Prairie des Femmes, for example, appears in a Spanish land document dated 1797 and on Barthélémy Lafon's 1806 map of the *Territoire d'Orléans*. Prairie Gros Chevreuil also appears on Lafon's map, while it and Prairie Laurent are mentioned in explorer and surveyor William Darby's 1818 handbook *The Emigrant's Guide to the Western and Southwestern States and Territories*.[2]

Paddling along, we soon arrived at Leonville [30.473495, -91.98005]. Interestingly, during Prohibition this area was known as a hotbed of moonshining. In fact, the entire triangular region between Leonville,

Arnaudville, and the nearby unincorporated community of Pecanière was well known for its bootlegging. This being said, I have found no evidence of rum-running on the Teche itself during Prohibition. The narrowness of the bayou would have afforded little if any secrecy and, thanks to the coming of modern roads, automobiles would have provided a faster, and more elusive, means of distribution.[3]

On the outskirts of Leonville we encountered enough suburban-style houses to remove the illusion of paddling through wilderness. We reached the town at 11:45 a.m. to the sounds of a church carillon (or at least it sounded to me like a carillon).

Just south of town, as my team and I headed back into the countryside, we spotted a nutria rat on the bank. With a flash of its pumpkin-orange teeth and long hairless tail, it scurried away to hide. Within the hour, just below the Oscar Rivette Bridge [30.447948, -91.924641], we saw a large owl. Five minutes later we passed under high-voltage transmission lines, whose audible hum and static made me nervous to sit between them and a body of water.

We observed pecan pickers here and there just north of the next bayou community, Arnaudville, as well as a large bird I listed as an anhinga. I'm no ornithologist, however, but it was large, sported dark plumage, and had black legs—perhaps not an anhinga, but a little blue heron? We would see many brilliant white herons, or egrets as they are also called, on our trips, often individual herons whose hunting we repeatedly interrupted, spurring them farther and farther downstream ahead of our canoes.

At 1:45 p.m. we arrived in Arnaudville [30.397596, -91.930761], our terminus for this stage of the trip. We put ashore at Myron's Maison de Manger (excellent burgers), right below the junction of the Teche and Bayou Fuselier.

On this side of the Teche and Fuselier sat the *vacherie* of my colonial-era ancestor, Lyons-born planter Gabriel Fuselier de la Claire. It's an extravagant name, but if he had been a big wheel in France I doubt he would have chosen to brave the south Louisiana frontier. With its heat, humidity, mosquitoes, alligators, pirates, and potential for slave and Native American revolts, the French and Spanish colony was no paradise despite its idealized portrayals in literature (à la *Evangeline*).

Preston and Jacques read a sign erected by the General Franklin Gardner Camp of the Sons of Confederate Veterans. It marks the reputed spot where in 1863 the Rebels scuttled several of their supply steamboats. Source: photograph by the author.

Fuselier de la Claire prospered in Louisiana, however, as did his heirs—but somewhere down the line, probably during Reconstruction, the Fuseliers lost their fortune and ended up marrying into my clan of subsistence-farming Cajuns. The same went for the de la Morandières, de Livaudaises, and other formerly affluent French Creoles who by twist of fate became my ancestors. A generation or so after marrying into the Bernards, little remained of their previous culture, though my late Cajun grandmother recalled her grandmother-in-law, the last de la Morandière of the line, speaking "that fancy French."

A folksy hand-painted sign hung beside the bayou near the landing at Myron's, erected by the local Sons of Confederate Veterans chapter (or "camp," as they call it). Placed for the benefit of boaters, the sign identified this as the spot where, in April 1863, Confederate forces fleeing General Banks's invaders scuttled the supply steamboats *Darby*, *Louise*, *Blue Hammock*, and *Crocket* (sometimes called the *Cricket*). How many bullets, swords, balls of shot, and other Civil War artifacts sat right there at the bottom of the bayou, only a few feet beneath our canoes?

Similar thoughts came to me repeatedly as I descended the Teche. I imagined bones, swords, cannons, revolvers, doubloons, and all sorts of other artifacts passing underneath me, perhaps within reach. Who knows what the mud and muck of the Teche is waiting to reveal?

After nineteen miles of paddling we were exhausted, and of the four of us I was by far the most miserable. My arms hurt as though I'd lifted barbells all day, and my palms stung with blisters and glistened with a metallic sheen from gripping the aluminum paddles. "Next time," I noted to myself, "bring gloves." By the time I reached home, I hurt all over. My bones seemed to ache to the marrow. No matter in what position I sat or lay, I could not ease the discomfort. This was my penalty for undertaking the trip without any physical preparation. I swallowed some aspirin and slept horribly, but by morning the pain had disappeared except for vague aching in my upper arms. Fortunately, this pain did not return after future stages. I needed this first stage to break myself in.

Stage 2: Arnaudville to Parks

Preston, Jacques, Keith, and I made stage two of our run down the Teche on November 5, 2011, putting in where we left off last time, at Arnaudville, right behind Myron's Maison de Manger. That day the sky was mostly clear and the temperature was 41°F in the morning, 72°F in the afternoon, and in the low to mid-50s°F by the time we got started at 10 a.m. We paddled southeast with the current toward our stopping point, the town of Parks.

The first town we reached that day was Cecilia [30.336093, -91.855681], known in colonial times as La Grande Pointe. There many houses lined the Teche, one of which had a dock sporting a ship's nameplate reading *Argosy President*. It sounded like a riverboat casino, many of which can be found along the Gulf Coast—not on the Teche, though there is a land-based casino along the bayou. (The nearest riverboat casino is the *Amelia Belle*, located on Bayou Boeuf about thirteen miles from the Teche's mouth.) Some sleuthing revealed that *Argosy President* was the former name of a 180-foot, 1,200-ton "platform ship" now known

We spotted this riveted pipe protruding from the bayou around Cecilia. I believe it once connected to a steam pump. Many of these pumps stood along the bayou in the late 1800s and early 1900s, when farmers used them in rice production. Source: photograph by the author.

This piling is typical of the red-brick ruins found at a number of sites along the bayou. Some of them, I believe, once supported steam pumps. Source: photograph by the author.

as the *Ensco President*. Given south Louisiana's deep-rooted ties to the petroleum industry, it didn't surprise me that the *Argosy President* had supplied offshore oil platforms.

We soon paddled under Poché Bridge [30.312121, -91.902679], known for the *boudin* (a spicy sausage) sold at nearby Poché's Market, a veritable culinary institution located on the west bank of the Teche. A short time later we spotted a patch of land occupied by numerous chickens, each dwelling in its own little coop. They looked to me like fighting cocks, even though Louisiana outlawed cockfighting in 2008. Some south Louisianians condemned the ban, asserting that cockfighting was a Cajun tradition and ought to be preserved.

It was here, around Cecilia, that we first saw signs of beavers, whose gnaw marks now and then decorated tree trunks along the bayou. Many of these marks appeared fresh, yet we never spotted any of the creatures at work. On a later stage of the trip, when we diverted onto Bayou La Chute for a short paddle, we saw a few beaver slides—paths carved in the mud by the semi-aquatic rodents as they glided into the water on the same spot repeatedly. But again, no actual beavers, which are largely nocturnal.

On this stretch I noted the many species of plant life along the Teche: live oak, cypress, pecan, reeds and rushes, elephant ears (an invasive species), Chinese tallow (another invasive species). Occasionally we saw clumps of cypress knees rising from the water or from the muddy *batture*, the sometimes flooded strip of land between the bayou's bank and the water's edge. Some of the cypress knees along the Teche grow two or three feet in height.

It was on this section of the Teche that we found an interesting artifact protruding from the water: a rusted but intact riveted pipe [30.351333, -91.887117] about a foot in diameter and running eight to ten feet before disappearing into the dark, soggy bank. It looked like a smokestack from a steamboat. There were certainly plenty of steamboats on the Teche from around 1830 to the steam era's final gasp around World War II.

I suspended judgment until I collected more evidence and eventually I believe I discovered the object's purpose. Paddling downstream over several months, we saw here and there what at first appeared to be tombs, inevitably made of the red-orange brick common to the region. Some of these "tombs" sat close to the water's edge; others crumbled into the water and disappeared beneath the bayou's surface. These brick "tombs," however, often sprouted large iron bolts clearly intended to hold large pieces of machinery in place.

At first I thought these "pilings," as I began to call them in preference to "tombs," must have been associated with the sugarhouses that once lined the bayou from St. Martin Parish to its mouth. Then we noticed a piling with rusty iron machinery still bolted to it. It turned out to be a steam pump. We stopped the canoe so that I could examine the artifact more closely. Letters embossed in the oxidized cast iron read:

IVENS DOUBLE SUCTION PUMP
BOLAND & GSCHWIND CO LTD
NEW ORLEANS LA

Rising vertically from this steam pump stood a riveted iron pipe very similar to the one we saw in the bayou near Cecilia.

But what was the purpose of this pump and the three or four others we eventually found along the Teche? Did they pump water from the Teche, or into the Teche?

Sugarcane planting required excellent drainage, as historians Glenn R. Conrad and Ray F. Lucas observed in *White Gold: A Brief History of the Louisiana Sugar Industry, 1795–1995*: "After the introduction of steam power, many planters installed steam water pumps to pump water over the protection levees into the back lowlands." But the steam pumps we saw were not on the back lowlands. On the contrary, they stood on the headlands near the bayou.[4]

It now seems likely these pumps were used to flood rice fields, because from the late nineteenth century into the early twentieth century many famers along the Teche actually grew rice. It is not by chance that the Conrad Rice Mill (maker of Konriko products) stands in New Iberia, just blocks from the Teche. The mill's location is a vestige of these nearly forgotten rice-growing days along the bayou.

Preston, Jacques, Keith, and I continued downstream on the Teche. South of Cecilia the bayou skirted Interstate 10, the superhighway stretching from Florida to California. The sound of speeding vehicles became increasingly loud—an unavoidable incursion of the rapid modern world into our slow exploration of the Teche. To compound this feeling, a B-52 bomber circled overhead, creating mammoth shadows on the clouds below it. I'd never seen a B-52 except in photos until the previous weekend, when I attended an airshow in Lafayette. Now I'd seen two B-52s in only a week.

At this time we encountered our first motorboat on the bayou. Previously we saw only canoes, pirogues, and other small wooden boats, all paddle-driven, and none of them actually on the bayou, but on shore, leaning against sheds or sitting upside down on blocks. Even the motorboat in question sat moored to a dock. It drove home the point that the Teche so far seemed underutilized by sportsmen and pleasure boaters. We had traveled some thirty miles from the bayou's headwaters and had yet to see any other boaters. Perhaps it was the time of year.

At 1:40 p.m. we passed under I-10 [30.290735, -91.928587] and reached the outskirts of Breaux Bridge. From this point on we saw increasing numbers of bushlines, trotlines (or troutlines, as some say), and juglines, all simple but effective means of fishing on the Teche and other waterways. In the case of a bushline, a fishing line with hook and sinker is tied to a low-hanging branch and left unattended to make a catch. A trotline, on the other hand, consists of a line tied horizontally over the water, say from one low branch to another, and from which hang several fishing lines with hooks and sinkers. A jugline consists of a floating plastic jug (often an empty soda or bleach bottle), to which are tied a hook and sinker, as well as a weight to keep the jug from drifting. It is understood that one never touches another's lines, but once I inspected a jugline out of curiosity. Pulling up the line I found it baited with chicken and weighted by a railroad spike. I immediately put the jugline back in the water so that it could "do its thing."

We shortly entered Breaux Bridge [30.275719, -91.897652], a town that traces its origins to the late 1700s, when an Acadian exile named Firmin Breaux built a small footbridge across the bayou. Back outside town we passed the entrance of the Ruth Canal [30.245531, -91.8811] in the west bank. Running to Lake Martin and from there to the Vermilion River, this canal had been excavated around 1920 for the purpose of diverting water to the Vermilion for rice farming. Here we spotted a blue heron on the hunt and two or three dead animals in the water—deer or goats or large dogs, I couldn't tell. We saw enough of these in the Teche to dissuade me from wanting to swim in the bayou, much less to drink water from it or eat fish from it. I admire, however, the brave appetites of those bushline, trotline, and jugline fishermen. (Since writing the above, I have indeed eaten a traditional Cajun *courtbouillon* made with Bayou Teche catfish—and suffered no immediate ill effects.)

Toward the end of the day we encountered our first fellow boater.

My canoe team and I landed at Cecile Rousseau Memorial Park [30.214696, -91.82829] in Parks at 4:45 p.m., having covered 23.6 miles in 5 hours, 57 minutes.

Stage 3: Parks to Loreauville

The third leg of our Teche trip took place on December 3, 2011, and ran between the towns of Parks and Loreauville. The temperature that day was 55°F in the morning, 81°F in the afternoon; the sky was partly cloudy. Present, as always, were Keith Guidry and myself. On this particular stage we were accompanied by Donald Arceneaux, who shared our interest in the bayou and its history. Donald would follow us in his small one-man kayak.

We put in at 9 a.m. at Cecile Rousseau Memorial Park in Parks, which sits in the stretch of Teche known in colonial times as La Pointe de Repos ("Point of Rest"). It was here that many Acadian exiles settled in the eighteenth century after first settling farther downstream for a time.

As we left Parks, we passed the site of Promised Land [approx. 30.206166, -91.826005], the reputed slave cemetery that anthropologists and local citizens hope to document and preserve as a historical landmark.

Canoeing by several modern houses, we soon reached the community of Levert-St. John [30.158533, -91.812519], the first sugar plantation we encountered (to my knowledge) since starting our trip at Port Barre weeks earlier. Operating under the modern corporate name Levert-St. John LLC, the plantation is owned by the Levert Companies (though the mill itself is owned by the Louisiana Sugar Cane Co-op). As the

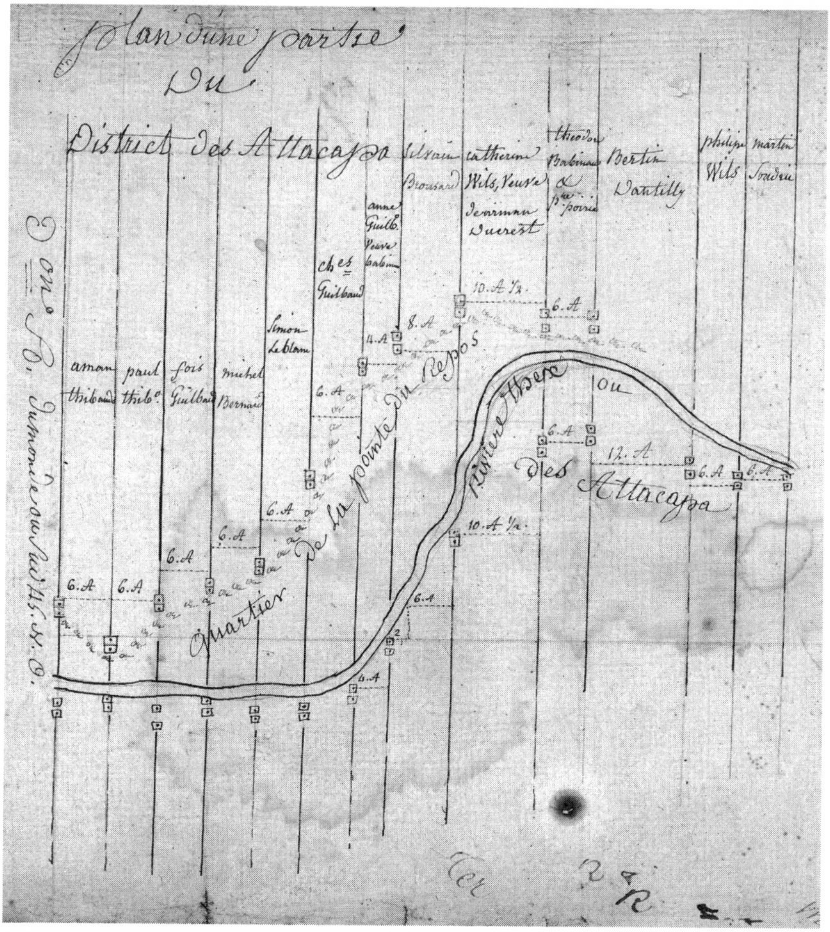

A late-18th-century land ownership map of the Teche around present-day Parks. Note it refers to the area as the "Quartier de la Pointe du Repos" and calls Bayou Teche by two names, "Rivière Thex or [Rivière] des Attakapas." Source: Library of Congress, Washington, DC.

firm's website noted: "The principal asset of this company [Levert-St. John] is agricultural land (approximately 11,000 total acres), all in St. Martin Parish, Louisiana. The Levert-St. John LLC lands are made up primarily of four plantations, St. John, Banker, Burton and Stella, all in St. Martin Parish." Of that 11,000 acres, approximately two-thirds is dedicated to sugar production, the remainder focusing on crawfish and rice production.[5]

The Louisiana Sugar Cane Co-op mill, still referred to by many as the Levert-St. John mill, located just upstream from St. Martinville. The now-defunct swing bridge bears the date 1895. Source: photograph by the author.

The sugar refinery at Levert-St. John dominates the surrounding cane fields. We were fortunate to canoe past the structure at the height of grinding season (October to January), when the complex belches enormous steam clouds into the winter sky. At any given time sugar mills either smell pleasantly sweet, like molasses, or they emit a god-awful stench, to which nothing polite compares. Levert-St. John must have smelled decent that day, because I made no reference in my journal to the odor.

In addition to the hissing, breathing refinery, we saw here the now defunct 264-foot-long, 14.7-foot-wide Levert-St. John Bridge, which appears on the National Register of Historic Places (site #98000268, added 1998). A double Warren truss swing bridge made of rusted steel girders, it was constructed in 1895 for railroad and pedestrian traffic. Eventually, however, the rails were covered with asphalt to form a single-lane bridge for automobiles (which I remember driving across in the 1980s—not the most secure experience).

After paddling around the bridge, and underneath it to inspect its swing mechanisms, Keith, Donald, and I continued toward St. Martinville. On the outskirts of town we reached the Longfellow-Evangeline State Historic Site [30.135863, -91.819224], the centerpiece of which is Maison Olivier, a "Raised Creole Cottage . . . which shows a mixture of Creole, Caribbean, and French influences." The home was built circa 1815 by sugar planter Pierre Olivier Duclozel de Vezin, from whom two of my canoeing partners, Preston and brother Ben, traced a direct lineage. I could not see the cottage from the bayou, but I did see Spanish longhorn cattle—a fitting living history display, because Spanish longhorn was indeed the variety of cattle raised in Louisiana during its colonial period. This variety is no longer seen in south Louisiana (at least not from my experience), although I have seen for myself longhorns still inhabiting parts of the Texas Hill Country.[6]

Immediately south of Longfellow-Evangeline, and just beside St. Martinville Senior High School (Preston and Ben's alma mater), one can see behind a low dam the Cypress Bayou Coulée Canal [30.131748, -91.822636] running northwest into Bayou Tortue Swamp. A major wetlands area, Bayou Tortue Swamp begins on the north near Breaux Bridge and stretches to the Terrace Highway (Highway 96) on the south, and from the natural Teche levee on the east to an area abutting the Lafayette Regional Airport on the west. I grew up in Lafayette never knowing such a large swamp existed nearby.

As we neared downtown St. Martinville, Keith showed me two peculiar landmarks. One of these was a boat made of cement, now dilapidated and slowly sinking into the bayou. Cement boats, it turned out, were not unheard of. The other landmark was the site of a small cave [30.124257, -91.826246] where local children used to play. Infinitely smaller than the one in which Tom Sawyer and Becky Thatcher became lost, the cave entrance has now been sealed with cement for reasons of safety; but you can see the cement patch over the cave's entrance, about a hundred yards above the downtown bridge over the Teche. Decades ago I explored a similar cave in a bluff of the Vermilion River, just south of the bridge over which E. Broussard Road (Highway 733) crosses that bayou. There was not much to the cave: its pale clay walls extended into the bluff about ten or twelve feet before ending in a small hole in the ceiling, which in turn opened onto a pasture.

From the bayou at St. Martinville we saw St. Martin de Tours Catholic Church, the Acadian Memorial, the Museum of the Acadian Memorial, the adjacent African-American Museum, the Old Castillo Hotel, and the Evangeline Oak [30.122102, -91.827457]. It was under that oak the Acadian maiden Evangeline went insane pining for her lost love Gabriel . . . well, that is how I heard the story.

American poet Henry Wadsworth Longfellow mentions St. Martinville in his epic poem *Evangeline*, first published in 1847: "On the banks of the Têche, are the towns of St. Maur and St. Martin. / There the long-wandering bride shall be given again to her bridegroom, / There the long-absent pastor regain his flock and his sheepfold."

In his study *In Search of Evangeline*, historian Carl A. Brasseaux has demonstrated that Evangeline never existed, nor did Emmeline Labiche, the reputed "real life" Acadian exile on whom Longfellow based the character of Evangeline. Thus, neither Evangeline nor Emmeline are buried in "Evangeline's Tomb" next to St. Martin de Tours Church. Yet the events against which Longfellow set *Evangeline*—the British military's brutal eighteenth-century expulsion of the Acadians from Nova Scotia and the eventual arrival of some of those exiled Acadians on the Teche—are factual.

As we left St. Martinville and paddled into the countryside, I spotted a few ibises on a mud flat, pecking away at minnows. They were young ibises whose plumage had not yet turned completely white. As a historian I found these birds of interest because the ibis, with its distinctive downturned beak, is strongly linked to ancient Egypt. There religious belief associated the ibis with Thoth, the ibis-headed god of wisdom and writing. According to Herodotus, the father of history, anyone who killed an ibis in Egypt, whether on purpose or accident, was put to death.[7]

A little after 1 p.m. we spotted the Keystone Lock and Dam [30.071095, -91.828859], a convenient line of demarcation between the upper and lower Teche. Near the dam floated several islands of water hyacinths, the largest measuring about seventy-five feet in length. Since starting our trip at Port Barre, we had worried about hyacinths. An invasive

Named after a local sugar plantation, the Keystone Lock and Dam was built in 1913 to raise the water level of the upper Teche. Source: photograph by the author.

species, we feared it would at some point block our progress, clogging the entire bayou from one bank to the other. Those we now saw were not enough to stop us; we merely paddled around the aquatic plants, or plowed straight through them, feeling their drag on the canoe. Aerial photographs showed even denser masses of water hyacinths ahead of us, near Baldwin and Franklin. But we would not reach those towns for a few more stages.

As we closed in on the dam, the sound of water topping the cement structure became louder. Here we had no choice but to do a portage. I don't know how people pronounce *portage* elsewhere in the country, but in south Louisiana the most common pronunciation I hear is the French pronunciation, which is POR-TAHZH. In fact, it was the only pronunciation used by my canoeing team, and it is how everyone in the region pronounces the word when saying the name of a nearby waterway, Bayou Portage. What is perplexing, however, is that the French place name Fausse Pointe—an archaic name for our stopping point that day, Loreauville, as well as for the entire stretch of the oxbow on which Loreauville sits—is almost universally pronounced (that is, mispronounced) as FAW-SEE POINT instead of FAHS POINT. (The term is used most often today in reference to Lake Fausse Pointe and the state park that borders it—but I refer to earlier uses of Fausse Pointe.) This Anglicized pronunciation jars me, just as the English pronunciation of *portage* jars me. Interestingly, no one (not even those who correctly pronounce *Fausse*) bothers to pronounce *Pointe* in the French manner (PWANT). I myself am guilty of this mispronunciation.

However you choose to pronounce it, we made the portage, hauling our canoe and Donald's kayak up a steep embankment on the east side of the Teche, carrying the boats about a hundred yards along a trail through a patch of woods, lowering them down another embankment (much steeper than the first), and putting them back into the water just below the lock and dam. Here the Tour du Teche race organizers had sunk some 4x4 posts in the embankment and, evidently running ropes between the posts, had rigged a method for lowering canoes into the bayou. Unfortunately, no ropes were present on our visit, so we had to make do. (Were we trespassing here? It was unclear: we exited the bayou on federal property, but at some point may have entered private property. I say this because just before we lowered our canoes back into the water we spotted a "Private Property" sign—but, confusingly, it was facing the wrong direction, so that canoers saw only the back of the sign until it was too late. I leave this matter for others to resolve.)

Once back in the water, we continued to paddle downstream, noting juglines on the bayou and beaver gnaw marks on the tree trunks.

On this stretch of the bayou not one, but two watermills operated in the early nineteenth century. Grinding sugarcane cut from nearby fields by slave labor, these mills relied not on the sluggish current of the Teche for power, but on currents produced by manmade canals. These canals ran from Spanish Lake to the Teche, whose surface sat many feet below that of the lake. As a result, the water from Spanish Lake essentially ran downhill to the bayou, creating enough force—probably augmented by a dam on the mill sites—to turn waterwheels attached to grindstones.

One of these canals stretched from Spanish Lake to the Teche through the DeBlanc plantation, later called Keystone Plantation after its purchase by a Pennsylvanian. One can see on old plat maps the canal labeled "mill race" (*race* in this sense meaning a narrow water channel used for industrial purposes). Though not much more than a ditch and long abandoned to the wild, this canal still exists and is known as Keystone Canal. As a superintendent of Keystone Plantation, John Peters, stated in 1899, "Many years ago [1845 or earlier] the water in the [Keystone] canal was carried by a flume down near the sugar house and was utilized to run a water wheel near the Teche for a sawmill and to grind corn and to even grind cane at the sugar house."[8]

The other canal ran from Spanish Lake, or from a "slough" adjoining Spanish Lake, to the Teche through the St. Marc-Darby plantation, later called the Wyche or Belmont Plantation. The ruins of this watermill can still be seen at Belmont: thousands of red-orange bricks strewn about the overgrown banks of the canal, here arranged into a sturdy wall, there stacked into small, crumbling, vaulted enclosures, which perhaps were furnaces for boiling cast-iron pans of sugary extract. Over two hundred years old, the site has not yet been formally examined by archaeologists or historians. It may be that similar watermill ruins await examination along the Teche at Keystone.

While discussing this stretch of the Teche, I should return to Longfellow's *Evangeline* and its line reading "On the banks of the Têche, are the towns of St. Maur and St. Martin." This reference to "St. Maur" always puzzled me, because today there is no St. Maur along the Teche. Nor did I know where it might have stood. Granted, the source is a poem, but Longfellow rooted *Evangeline* in fact, having researched south Louisiana from a distance while writing his epic. There was a medieval monk known as "St. Maur," and there are several communities in France named "St.-Maur."

I have to credit Donald with solving this mystery. He pointed out that prior to *Evangeline* the phrase "St. Maur" appeared in only one known source regarding the Teche Country. In the aforementioned 1818 book *The Emigrant's Guide*, William Darby referred to "Madame St. Maur's" homestead as sitting on the Teche between New Iberia and St. Martinville. Elsewhere in the book he mentioned the Teche flowing downstream "to M. St. Maur's plantation, where commences the Fausse Point[e] bend." He then referred to "M. St. Maur's house." Finally, many pages later he mentioned "Mad. St. Maur's" where "the sugarcane still endures the vicissitudes of the climate."

Longfellow evidently consulted Darby's book as a source for *Evangeline*, using the term "St. Maur" not as the name of a landowner or plantation (as Darby does), but as the name of a town along the Teche. Here, however, was where Donald made his most vital contribution to unraveling the mystery: armed with Darby's reference to "Madame St. Maur" and his description of the St. Maur homestead as sitting on the Teche "where commences the Fausse Point[e] bend," Donald concluded that Darby meant not "St. Maur" but "St. Marc"—that is, the widow Madame St. Marc-Darby (no known relation to William Darby), who did indeed operate the St. Marc-Darby sugar plantation at the head of the Fausse Pointe oxbow on Bayou Teche. In short, "St. Maur" was a mangling of "St. Marc"—a mere spelling error by Darby or his typesetter that Longfellow picked up and perpetuated in *Evangeline*.[9]

So much for the lost village of St. Maur.

Donald, Keith, and I passed the entrance to the Joe Daigre Canal [30.064878, -91.826166], which runs a short distance to Bayou Tortue. (Daigre was a mayor of New Iberia in the early to mid-twentieth century. He also has a bridge over the Teche named for him in New Iberia.) In turn, that bayou wraps itself around the northern bank of Spanish Lake before entering Bayou Tortue Swamp.

At 2:50 p.m. we reached Daspit Bridge [30.040546, -91.800843], named for a small unincorporated community on the outskirts of New Iberia. Around this time we noticed a sweet molasses smell—another sugar refinery in operation. We passed many modern suburban houses on the east bank and at 3:20 p.m. I spotted a wall of green in front of us—What was it? I wondered. A painted green wall, perhaps the side of a large barn? It seemed to rise two stories or more. I then realized it was a grassy hill rising above the bayou's bluff. (An archaeology friend later suggested the hill as a likely candidate for an Indian mound.) "What is this place?" I asked Donald, who, it turned out, knew precisely what it was. He'd scouted ahead, putting his kayak in near Daspit Bridge a few days earlier. Now he was toying with me, awaiting my reaction to the sight of the grassy hill, below which a *coulée* or small bayou entered the Teche.

The mouth of Bayou La Chute, where in the 1700s a ten-foot waterfall emptied into the Teche. Source: photograph by the author.

It was the mouth of Bayou La Chute [30.03667, -91.786284] and the site of the eighteenth-century waterfall that emptied into the Teche.

Donald had identified this site as the probable location of the waterfall—it just made sense, topographically. Additional research on my part revealed that locals called this small tributary Bayou La Chute . . . and "La Chute" in French means "the waterfall." There is no waterfall today, its ten-foot drop, made of dirt or clay, having been removed by human activity or eroding away naturally.

A paddle up Bayou La Chute is worth the effort. Although modern neighborhoods encroach on more than one side (including a neighborhood under construction), I found Bayou La Chute in some ways more beautiful than the Teche. Because of its dark swampiness and winding narrowness, lined with bamboo (a nonindigenous plant no doubt loosed from someone's garden), it reminded me of an amusement park jungle ride.

Stopping at a fork in Bayou La Chute, we turned around and made our way back to the Teche and continued toward Loreauville. With the sun sinking and the sky dimming, we had one more discovery to make that day, the so-called "tombs" [30.062517, -91.769417] I mentioned previously. As I stated, I no longer think these are tombs; rather, I believe they are pilings that once supported steam pumps, and that these pumps were used in the days when many Teche Country planters cultivated rice instead of sugar.

But rice cultivation had faded into oblivion, and now we were once more firmly in sugar country, passing through one old sugar plantation after another—some now transformed into neighborhoods, others still producing sugar, but, of course, without the slaves who once carried out the grueling tasks associated with making sugar. These included cutting cane by hand, hauling it to the sugar house, feeding it into the rollers, boiling the saccharine juice, refining it again and again until, at the exact moment, a "strike" was poured, creating granulated brown sugar seeped with molasses. Loaded into oversized barrels called hogsheads, the commodity traveled via Teche steamboats to New Orleans, where it would be sold on the local market or transferred to schooners for sale on the East Coast.

On the outskirts of Loreauville we saw three immense boat-building facilities and a couple of vessels associated with the petroleum industry—a quarter barge and a spud barge. These sights reminded us that although we paddled inland on a bayou whose mouth did not empty into open water, we were nonetheless not far from the Gulf of Mexico.

Modern homes and businesses now ran along the bayou at Loreauville. We drew up to our designated landing, the town's bridge [30.056791, -91.740185], which stands next to a volunteer fire station and an old jailhouse. Some locals hope to convert the fire station site into a park, and the jailhouse into a museum dedicated to local history, particularly the 1765 settlement of Acadian exiles along this stretch of the bayou. Indeed, archaeologist Mark A. Rees of the University of Louisiana at Lafayette has searched this section of the Teche for artifacts left by those exiles; and his search continues today.

We debarked about 5:30 p.m. and hauled the canoe and kayak up a high bluff below the jailhouse. Donald had an easier time with his small kayak, but Keith and I struggled with the seventeen-foot metal canoe. Wrestling with the boat, the two of us suffered simultaneous asthma attacks—a humorous coincidence, but only in retrospect. Fortunately, we each brought an inhaler, and soon we were enjoying hamburgers at the nearby (and aptly named) Teche Inn, located a block or two from the bayou. Distance covered that day: 19.15 miles in approximately 8.5 hours.

Stage 4: Loreauville to Jeanerette

The fourth stage of our trip down the Teche occurred on January 29, 2012, and covered the section between Loreauville and Jeanerette. Keith Guidry, his son Ben, and Jacques Doucet accompanied me. The temperature that morning was a brisk 42°F; in the afternoon it reached a comfortable 63°F. We put in at 8:50 a.m. and within five minutes noticed one of the area's several boat-building facilities. These are large boats, perhaps eighty to a hundred feet in length. I reminded myself again that the open waters of the Gulf of Mexico lay fairly close, where such boats would not seem as out of place as they did here.

Our starting point, Loreauville, sat halfway down the meander of the Fausse Pointe oxbow. This meander bulged eastward toward the Atchafalaya swamp before shooting back toward New Iberia. Tackling the oxbow's lower half, we paddled past the mouth of Teche Lake Canal (also known as the Loreauville Canal) [30.023922, -91.731677], located in the east bank about 2.5 miles south of Loreauville. Despite its name, Teche Lake Canal doesn't lead to "Teche Lake"—there is no such lake—but to Lake Fausse Pointe.

Still on the oxbow, we next passed through Belle Place [30.013368, -91.729617], which took its name from an old sugarcane plantation. ("Belle Place" lives on in the name of a local middle school.) It was in this unincorporated community, near its present-day bridge over the Teche, that sugar planter Euzebe Verret, after tending his fields, would sit "under a giant oak tree overlooking Bayou Teche . . . cracking and eating pecans, while dispensing sound advice to passersby. The young men of the countryside would gather around to sit in a circle around his feet and listen to his stories and jokes"—no doubt all told in French. "Others," we are told, "played mumblety-peg [a knife-tossing game] nearby in the dusty earth with several friendly bets hinging upon the outcome of the game." This tradition may have ended with Verret's death in 1936 or, if it lingered, perhaps stopped with the coming of World War II, when the young men of the parish went off to join the conflict.[10]

Next on our trip that day came Morbihan [30.015858, -91.775107], another unincorporated community deriving its name from an old sugarcane plantation. Morbihan sat halfway between Belle Place and the city limits of New Iberia. Here, only about 1,400 feet from the bayou, stands a cement obelisk [30.012233, -91.772123], which I would estimate to be about fifteen feet in height, dedicated to the memory of W. Knighton Bloom (1866–1934), an out-of-state Congregationalist minister. Although white, Bloom earned the admiration of local African Americans by establishing a church facility on this site in the early twentieth century. It was dubbed Kamp Knighton [*sic*]. The camp existed for the enrichment of African American youths, not only from Morbihan but from throughout Louisiana, Arkansas, and Texas. A public playground, Morbihan Park, sits on the site today, but the lane beside the obelisk bears the name Camp Knighton Road.

Between Belle Place and Morbihan sits the Cajun Sugar Co-op, another refinery that converts the region's raw sugarcane into commercial-grade sugar. A bit closer to New Iberia, however, sits the rusting detritus of a less fortunate refinery, the Iberia Sugar Cooperative. It went defunct around 2005. I miss this refinery, a little: in winter the molasses-sweet scent of raw sugar floated over my New Iberia home, while the mill hissed and murmured in the distance. Sometimes, however, the scent became sour—but no one complained much: the stench lingered only a day or two, and then the sweetness returned, or there was no scent at all.

It was on this part of the Teche, just upstream from New Iberia, that U.S. Navy agent and surveyor James Cathcart witnessed the following scene in 1819: "[W]e embark'd [from New Iberia], and steer'd up the Teche, for St. Martinsville, & soon after met two canoes with a large family of Indians in them, viz. [that is]: 4 women, one of whom was half white, as many men & several children, besides a large dog couchant [lying down with head up] in the bow, & their baggage. One white man would have overturn'd either of those canoes if not extremely careful. . . ."

The haze that gives this image its allure was actually smoke from a pile of burning tires. A careful eye can detect a PVC pipe emptying into the bayou as well as a full-size plastic deer decoy. Source: photograph by the author.

We reached New Iberia at 11:50 a.m., more or less, paddling around the sharp bend in the bayou known in earlier times as Petite Fausse Pointe [30.015384, -91.821025]. Near downtown—a vibrant small-town Main Street, full of shops and restaurants—a venerable red-orange brick structure, the Lutzenberger Foundry [30.012075, -91.821515], came into view. Established in 1871, the foundry cast parts for steamboats and sugar mills. The extant brick structure, however, dates from a bit later, around 1880.

Intriguingly, the Spanish colonial site of Nueva Iberia stood along the Teche in the vicinity of the foundry. In late 2012 I convinced my friend, archaeologist David Palmer of the University of Louisiana at Lafayette, to undertake a small excavation of the foundry grounds. Our goal: to search for signs of Nueva Iberia. He and I, along with a few volunteers, set up screens and dug shovel test pits. We uncovered lots of slag associated with the foundry, but no colonial-era artifacts. Not even a single glass bead, much less a conquistador helmet, or a swivel gun, or a cache of silver *reales*. I was disappointed; but David was pleased because he, unlike me, found the nineteenth-century artifacts of interest in themselves.

Passing along downtown New Iberia, which sat on the west bank of the Teche, we glided under the Bridge Street (Duperier Avenue) Bridge. It was here in the early twentieth century that bridge keeper Everard Viator and his wife—urged by the town's esteemed family physician, Dr. George Sabatier—christened each of their children with a name borrowed from Shakespeare. And so this spot on the Teche became home to Brutus, Cassius, Julius Caesar, Marc Antony, and Cleopatra Viator.[11]

On the downstream side of the Bridge Street Bridge we stopped at architect Paul J. Allain's boat dock [30.005253, -91.81603], which is open to the general boating public. Rising from the murky water near the dock, a row of pilings guards the submerged ruins of a nineteenth-century steamboat. As the local *Daily Iberian* newspaper reported in 2007, "A sunken ship that wrecked nearly 140 years ago was unearthed last year during a site excavation by New Iberia architect Paul Allain. . . . The vessel was buried below 4 feet of mud under the bed of Bayou Teche." A formal archaeological survey tentatively identified the vessel as the aptly named *Teche*, also called the *Tensas*. Parts of the vessel have been excavated, including a section of the keel.[12]

It was about this same stretch of the Teche that one venerable resident of New Iberia recalled in 2006:

> We did a lot of swimming in the bayou. The paddlewheel steamers would cruise down the Bayou Teche and some of the boys had pirogues that would follow after the paddlewheel steamers and hope that the wash from the paddlewheel steamers would turn their pirogues over. . . . And also in those days there were log trains in Bayou Teche, tugboats pulling logs of cypress from the swamps, the last of the cypress because the cypress swamps were dying out by that time. . . . And some of the boys used to ride their pirogues and . . . try to ride on the logs as they went along.[13]

Waved on by my wife, children, and some neighborhood kids, my canoe team and I paddled along New Iberia's City Park toward the Lewis Street Bridge. Heading back out of town, we passed under the fractured smokestack [29.9974, -91.800309] of the long-shuttered Charles Boldt

Paper Mill, which in the early twentieth century made corrugated fiber boxes from rice straw, a byproduct of the area's once booming rice production. There in the west bank we spotted the mouth of Nelson Canal [29.988465, -91.780083]. Here in 1863 Union and Rebel forces skirmished, as David C. Edmonds recorded in *Yankee Autumn in Acadiana*:

> During his observations along the Teche, Mexican war hero [Colonel William G.] Vincent, cunning as ever, found an opportunity too good to pass up. About two miles south of New Iberia, on the plantation of a wealthy New Orleans businessman and planter named S. O. Nelson, the roadway struck a deep hedge-lined drainage ditch (Nelson's Canal) which emptied the excess waters of the adjacent prairies into Bayou Teche. The only passage over the ditch was on a narrow brick culvert. A few yards upstream from that point, near the banks of the bayou, stood a thicket of cypress trees. What better place to conceal bushwackers, reasoned Vincent. . . . Though some details are missing, it appears that Robinson's "Louisiana" Yankees, together with Colonel Edmund Davis' 1st Texas Cavalry (Union), rode right into the trap. "Colonel Vincent ambuscaded them at Nelson's Bridge," wrote General Mouton that night, "leaving the road full of dead and wounded." Nonetheless it was but a short affair. Within moments the famed 2nd Massachusetts Battery of Light Artillery (Nims' Battery) joined the skirmish, driving off the Rebels with a barrage of screeching Schenkl shells.[14]

Below New Iberia a long, unbending stretch of the Teche runs through the countryside. So unbending, in fact, that a published history of nearby Jeanerette, our terminus that day, is aptly titled *Where the Bayou Runs Straight*. To the immediate west of this straightaway, and running parallel to it, sits the long expanse known to locals traditionally as Île aux Cannes (Island of Canes). Slightly downstream of it stretches the swath of farmland (much of it now made up of rural neighborhoods along Highway 90) once known unflatteringly as La Côte aux Puces (Flea Coast). The area is now, however, the Creole of Color enclave called Grand Marais (Big Marsh—though its French name is pronounced GRAND MAIR-REE in an Anglicized manner).

Around this time in our journey, Jacques came down with a migraine; he needed a Coke, he said—the only sure quick remedy. Keith took out his cell phone (he kept it in a Ziploc bag in a watertight tackle box) and called a friend who lived a couple miles downstream. At 2:47 p.m. we spotted Keith's acquaintance, along with son and dog, waiting in a boat to give Jacques a Coke. They sat in the shade just below the Louisiana State University's Iberia Research Station bridge [29.955195, -91.715308].

What a great picture, I thought, imagining myself a *National Geographic* photographer. "May I take your photo," I asked, pulling out my Nikon D60. "No," replied the man, muttering something about "the police."

We did not mention to Keith's friend that Jacques, in desperation for relief from his headache, had already put ashore at Olivier Bridge [29.979082, -91.7538] and purchased a Coke from a nearby convenience store—again shattering any illusion of "roughing it."

Incidentally, the LSU Iberia Research Station contains 900 acres of pasturage and 150 acres of plowland. The station produces "energy cane," which can be turned into biofuel, and crossbreeds cattle to make new varieties that thrive in the semitropics. It sits on the former site of Hope Plantation, a convict farm operated around the turn of the twentieth century by the Louisiana Department of Corrections. During World War II the station housed a German prisoner-of-war camp whose inmates were largely Afrika Korps troops.[15]

Keith's friend accompanied us down the bayou for a while and divulged that an abandoned smokestack [29.940404, -91.694351] ahead on the west bank had belonged to a lumber mill. This struck me as plausible, because the Teche and its environs, including the Atchafalaya Basin, had supported a vibrant cypress logging industry in the late nineteenth and early twentieth centuries.

Leaving Keith's friend, we entered an uninhabited stretch of the Teche. "Without houses," I recorded. "Woods both sides of bayou." The smokestack still loomed ahead and as we paddled closer I saw just below its apex a delta symbol, Δ—perhaps a clue to the identity of its former owner. (I later learned, however, that the Loisel Lumber Mill once stood on the site and, earlier, the Loisel Sugar Company.)

At 3:40 we reached Bayside [29.930642, -91.680479], the antebellum plantation home of nineteenth-century sugar planter F. D. Richardson. He penned a short but important memoir titled "The Teche Country

The smokestack at Loisel, site of former sugar and lumber mills, just upstream from Jeanerette.
Source: photograph by the author.

Fifty Years Ago," published in the March 1886 issue of the Confederate veterans magazine *Southern Bivouac*. In that memoir Richardson noted that years before Bayside's construction a "famous, tall, black live-oak stump" stood nearby, "'gloomy and peculiar,' up to 1800, and gave its name, *chicot noir* [literally "black stump"], to all the *arrondissement* [vicinity]." Richardson's daughter, Mary Louise, married former Confederate officer Dudley Avery of Petite Anse Island (now Avery Island, located some thirteen miles from Bayside), for whose descendants I myself work as historian and curator.[16]

Bayside marked our arrival in Jeanerette [29.918563, -91.666317], and at 4:07 p.m. we came to a stop at Jeanerette City Park on the west bank. That day we covered approximately 20.5 miles in 5 hours, 55 minutes.

Stage 5: Jeanerette to Baldwin

The fifth stage of our journey down the Teche took place on March 10, 2012, even though two of my canoers bailed at last minute. I canceled the trip; but around noon I called Keith—or did he call me?—to ask, "What if we go anyway?" We had never canoed without at least three people. It was a decent day for canoeing, however, and Keith and I both itched to go; so we went. We dropped off a vehicle at our stopping point, then backtracked to start the run at Jeanerette City Park. We put in at 1:30 p.m., much later in the day than we'd done previously. But Keith and I felt certain we could make the planned fourteen-mile trek before sundown.

The temperature peaked at 61°F that afternoon; an overcast sky hung overhead.

We paddled only a mile and a half before spotting a beautiful antebellum home [29.904153, -91.649329] perched on the west bank. "Albania?" I wrote in my journal. So it was. Built around 1840 by Charles Grevemberg—descendant of early Teche Country settler Jean-Baptiste Grevemberg dit Flamand—the dwelling is now the home of renowned artist Hunt Slonem.

Wisteria vines along the banks near Albania sprouted lavender flowers like bunches of grapes.

Beyond Jeanerette we paddled through the tiny community of Sorrel. Site of a former sugarcane plantation, the place is named for Jacques Joseph Sorrel, a French soldier who arrived in Louisiana around 1760 and who eventually owned thousands of acres of prime farmland along the Teche. Today a modern sugar mill sprawls atop the presumed site of the ancestral Sorrel home.

A short distance past Sorrel we spied the crumbling smokestack of the Adeline Plantation sugar mill. From a distance it is all that appears to remain of the once vibrant sugar-producing complex.

During this leg of the journey we began to see shrimp boats on the Teche. They would become a more common sight as we paddled closer to lower-Teche canals leading to the Gulf. Each shrimp boat bore a unique name, usually feminine. That day we encountered the *Myriah Lee*, the *Dixie Le* (which I assume was missing a vowel), the *Lady Flo*, and the *Miss Betty Ann* (which, though otherwise in excellent shape, appeared to have floated fifty or so feet inland during a spell of high water).

About 3:15 p.m. Keith and I reached the Sovereign Nation of the Chitimacha [29.89003, -91.526377], established in 1916, the same year the U.S. government officially recognized the tribe. For a distance both banks of the Teche ran through tribal lands. At one point we passed a man working on a moored boat: I recognized him as one of four Chitimachas with whom I was acquainted. He was, in fact, the current tribal chairman (the word *chief* having fallen out of favor). Keith and I stopped to talk with him for a few minutes. His sister, with whom I'm also acquainted, is one of the tribe's renowned basket weavers.[17]

Here is the town of Charenton, often identified in nineteenth-century sources as Indian Village, just as the curve on which it sits was often called Indian Bend. According to folk etymology, Charenton acquired its present-day name when a French immigrant, sugar planter Alexandre Frère, declared that anyone who chose to live on the mosquito-infested bend should be committed to the notorious insane asylum outside Paris called Charenton. How true this story is, I don't know, but the name has been associated with the place on Bayou Teche since at least 1849, when an advertisement in the Franklin newspaper referred to the "Charenton Post Office" at "Indian Bend." In their own language, the Chitimacha refer to this community today as *Caad hikutinki*.[18]

It was near this site, in 1779, that Spanish soldier Francisco Bouligny originally chose to locate Nueva Iberia, only to have the village swept away by floodwaters. This was also the spot described by Thomas Hutchins in 1784 as "the village de [chief] Soulier Rouge." In 1819 surveyor John Landreth (who called the place "Indian Beach") gave one of the earliest descriptions of the village, observing "a number of Indian cabins built nearly [neatly?] in a row. . . . These cabins have a neat light appearance covered and shut in entirely with the leaves of the palmetto, which keeps out the rain very well. Saw a number of children and young Indians running about in front of one cabin."[19]

In the graveyard of Immaculate Conception Catholic Church [29.885955, -91.52517], only about a block from where the Teche passes under Charenton Bridge, stands the tombstone of Clara Darden, master Chitimacha basket weaker. The memorial was erected in the early twentieth century by Mary Avery McIlhenny Bradford, daughter of Tabasco pepper sauce inventor E. McIlhenny of Petite Anse Island, and Neltje Blanchan De Graff Doubleday, of the Doubleday publishing family. Encouraged by Bradford and Doubleday, Darden, along with fellow weaver Christine Paul, labored to preserve the tribe's basket-weaving traditions—which do indeed survive to the present.

Canoeing deeper into tribal lands, Keith and I noticed that his GPS device showed us far off-course. According to it, we were paddling eastward into Grand Lake, which was not our intention at all. The device might be malfunctioning, we thought, but we were unsure. Earlier we confronted a fork in the bayou, and now we wondered if we chose the correct route.

As if by magic, a kiosk displaying a large map of the Teche appeared on the west bank [29.888629, -91.531006]. (Thank you, St. Mary Parish Cajun Coast Visitors & Convention Bureau!) As it turned out, we were precisely on course: the kiosk's map said so, unambiguously declaring "You Are Here"—and "Here" was the Chitimacha Boat Launch on Bayou Teche.

As Keith and I studied the map, two Chitimacha girls, with darkish skin, hair, and eyes, watched us intently. One looked about six years old, the other about four. The aspiring *National Geographic* photographer in me wanted to snap their picture, but that seemed inappropriate. I asked the girls if we still were on "the reservation" (a term I have since learned the tribe finds distasteful) and they nodded yes. I worried about them playing as they were, unsupervised, on the wooden pier, so I cautioned them, "Don't get close to the water or you might fall in." I disliked leaving them there, as did Keith, but we didn't know what to do other than warn them away from the water's edge.

Confirming that we were indeed on course, Keith and I soon spotted a monolithic sign rising on the west bank about a half-mile away. It read in huge luminous yellow letters:

C
A
S
I
N
O

It was Cypress Bayou Casino [29.872653, -91.537734], owned by the Chitimacha tribe. It seemed bizarre to me that only a few hundred yards away from our canoe stood a dazzling entertainment complex whose slot machines, blackjack tables, and roulette wheels never closed, and whose stage had featured, among numerous other celebrities, Lionel Richie, Dolly Parton, and Wayne Newton.

Still in the vicinity of Charenton, Keith and I canoed past a large cement structure [29.881543, -91.521699] on the east bank of the Teche. At first it reminded me of a World War II–era bunker or pillbox. As we canoed closer, however, I noticed wooden crossties atop the structure: it was a piling for a now missing railroad bridge. I thought perhaps the bridge had served a "dummy line," one of the narrow-gauge railroads that plantations once used to carry sugarcane from the fields to the mills. At nearby Patoutville [29.903721, -91.729066], for example, Enterprise Plantation had operated its own dummy line. In fact, it still has three vintage locomotives, the *Ida P.*, *Bessie*, and *Stephanie*, in storage. (Two other Patoutville locomotives, the *Lydia* and *Mary Ann*, underwent restoration years ago and since then have transported fun seekers around Six Flags Over Texas amusement park near Dallas-Fort Worth.) I checked an old quad map, however, and found that the bridge in question had belonged to the Missouri Pacific Railroad, whose line along the Teche had long since been ripped up.[20]

A few miles beyond Charenton we approached Baldwin, near which early Attakapas settler André Masse resided around 1770. Along this stretch of the bayou Keith and I noticed the ruin of an old wooden boat sitting in the overgrowth of the west bank. Despite its dilapidation, I could see that its builder had invested much time and skill in its creation. To make a boat entirely of wood, such as this one, is a lost art, or nearly so. I photographed the boat, took its GPS coordinates, and reported the find to Professor Ray Brassieur of the University of Louisiana at Lafayette. Professor Brassieur has a strong interest in south Louisiana's wooden boat-building traditions.

In response Brassieur observed that the remains indicated a "plank-on-frame" design with a "shallow model-hull (rounded hull), built of double-sawn frames." A tree fall, he noted, had severed the rounded fantail transom from the rest of the boat. He added, "It looks like the

Believed to be the remains of a motorized Louisiana fantail lugger from the period ca. 1915–50. Source: photograph by the author.

cylinder for the [propeller] shaft is exposed—it is likely mounted in the deadwood of the vessel. A piece of the *passe-avant* (narrow deck) remains attached to the port mid-section. I would expect to see the remains of a rear cabin on this vessel, but it was likely removed in the salvage of the engine, which is not present in the photo." In conclusion, Brassieur summarized, "I suggest that this is the remains of a motorized Louisiana fantail lugger. These vessels were built in greatest numbers during the period 1915–1935, but some continued to be built into the 1950s. I believe it is a significant artifact."[21]

Keith and I would leave it to others, if warranted, to determine ownership of the vessel and to decide whether or not to preserve and study its remains. We did try to remove a vintage bottle from the muddy bank near the vessel, but I lost my grip on the artifact and it disappeared into the murky water.

Toward the end of this stage Keith and I observed that the Teche had reversed its current, making our canoe noticeably more difficult to paddle. This reversal apparently stemmed from the outflow of the Charenton Drainage and Navigation Canal, the upper portion of which

[29.891257, -91.524274] extends northeast from the Teche oxbow at Charenton to connect with the Atchafalaya swamp. (We soon would reach the canal's non-contiguous lower portion, which runs southwest from Baldwin to the Gulf.) Above this canal the Teche flows downstream; below the canal the Teche flows upstream. Or so it seemed on that particular day. Regardless, our speed fell to 2 or 3 miles per hour, instead of our previous 4 or 5 miles per hour.

This decline might at first seem negligible, but it meant we spent about twice as much time as before covering a given distance. And this reversal of current, we speculated, would continue to work against us, reinforced by a similar outflowing of the Teche through the lower portion of the Charenton Drainage and Navigation Canal, and perhaps through the Calumet Cut if its west floodgate were open. Those channels lay only a short distance ahead. As a result, we decided to reduce the length of future stages to roughly the span of that day's stretch—about fifteen miles. Actually, the run that day totaled 14.1 miles. We covered it in 4.24 hours, reaching the Baldwin boat dock [29.833164, -91.542654] at about 6 p.m.

Stage 6: Baldwin to Franklin

We made the sixth leg of our Teche journey on June 3, 2012. The day was clear and bright, but miserably hot, reaching 92°F. The humidity was high, peaking at 90 percent and compounding our discomfort. Such conditions are common for semitropical south Louisiana in the summer.

We put in at 10:30 a.m. where we previously left off, the public boat ramp in Baldwin. Our terminus would be a tract of private land on the far side of Franklin. Keith and I paddled one canoe, while Keith's son Ben paddled a second by himself—a strategy that turned out to be a bad idea.

As we began that day's stage, Keith, Ben, and I noted pushboats and barges moored to the banks at Baldwin. Living in New Iberia only a few blocks from the bayou, I often hear pushboats blasting their baritone horns at the bridge keepers. I think of these boats and barges whenever I find it hard to imagine steamboats plying the bayou. Modern barges are roughly the same length and width as long-extinct river steamers and so give a good idea of what it must have been like to see one of those old giants on the Teche.

Not too long after departing Baldwin we became confused about which direction to steer our canoes. This uncertainty stemmed from a split in the bayou. Should we go right, following a clear, wide path leading into the lower portion of the Charenton Drainage and Navigation Canal [29.824757, -91.538485], otherwise known to locals as the Baldwin Cut (which runs to West Côte Blanche Bay and the Gulf of Mexico); or should we go straight, entering a silted-up, marshy channel not much wider than our canoes that clearly was a short remnant of an earlier course of the Teche?

We chose to avoid the silted-up channel, which seemed almost impassible, and instead headed into the Baldwin Cut. But now what? Always gregarious, Keith hailed some motor boaters to ask for directions. They told us if we paddled a short distance farther down the Baldwin Cut, we would find a channel to the left running back to the Teche proper. Taking their advice, we continued into the Baldwin Cut. An opening gradually appeared in the dense tree line to our left—a "detour" back to the bayou's true course.

Although we could not discern it at the time, we were in fact navigating around a wooded triangular island. The silted-up channel, the man-made Baldwin Cut, and the detour leading back to the Teche made up the three sides of the island. (I have since been informed that the name of this island, for unknown reason, is "Victory Island.")[22]

Around this time Keith and I noticed Ben falling far behind. Paddling his long aluminum canoe by himself, he simply wasn't able to keep up with us. So we stopped and tied Ben's canoe to ours, thinking we could tow him while he paddled just enough to ease the drag on our canoe. I don't think we understood the physics of the problem, because this plan did not work. After forty-five minutes of struggling we untied the canoes, and Keith and Ben switched places—Keith having more experience as a solo paddler. He also understood the nuances of the other canoe, whose keel was twisted slightly out of alignment. Only a deft hand could keep the boat from veering off to one side. The two canoes now clipped along at about three miles per hour, still slowed by the reversal of the bayou's current.

Mist hovers over the Teche the morning of one of our canoe trips. Source: photograph by the author.

Around this time we heard gunshots. About 150 yards ahead of us we saw the impact of bullets on the water. Keith shouted "Yo!" to let the unseen shooters know of our approach. It was merely some kids shooting a rifle for amusement. But there is something about being midstream in a canoe to make one feel like the proverbial "sitting duck." It was not the first time that day we'd encountered shooters. Less than an hour earlier we'd paddled up on two or three shooters on the bank. In that instance, however, they were aiming at targets on shore, not at the bayou itself. Except for Keith's Bowie knife and Ben's utility ax—the latter of which, as will be seen, ended up at the bottom of the Teche—we ourselves did not carry weapons.

We soon entered the head of the sizable oxbow known as Irish Bend [29.823072, -91.478546], so named because three early sugar planters who settled on the bend claimed Irish ancestry. It was here that Union and Confederate forces clashed at the Battle of Irish Bend (April 1863). Hoping to trap Confederates already engaged with Union troops farther down the bayou at Fort Bisland (present-day Calumet [29.698789, -91.373146]), U.S. General Cuvier Grover landed over 8,000 soldiers by steamboat on the western edge of the Atchafalaya. From there he marched them to the Teche. Waiting on the oxbow were a significantly smaller number of Rebels under Generals Richard Taylor and Alfred Mouton, who managed to stall the Union troops long enough for the main Rebel army to escape to fight another day.

Canoeing around the edge of the battlefield—privately owned farmland that sprouts sugarcane, just as it did at the time of the battle—we noticed a grassy mound rising from the west bank. A woman stood nearby on a pier; I called out to her and asked if the knoll were an Indian mound. No, she replied, pointing to a crescent-shaped cut in the bank, "It's the spoil from this old plantation slip." We had, she explained, reached Camperdown Plantation [29.844097, -91.489009], and its sugarhouse once stood beside the slip, opposite the mound of spoil. The slip had been used by vessels engaged in the sugar trade. Several of these crescent-shaped cuts appear along the Teche starting around St. Martinville, the bayou's upper limit for large steamboats. Called "turning basins," steamboats used them to turn around because the long vessels could not otherwise change direction on the narrow bayou.

Rounding the oxbow, we spotted a few rafts of water hyacinth. Keith had heard stories about Teche canoers having to slide into the water and drag their boats by rope through the tangled mats of hyacinths. Based on recent aerial photographs, we expected to run into nearly impenetrable clusters of the plant on this particular stretch of the Teche. Indeed, those photos made it appear that hyacinths clogged the bayou from bank to bank for several miles. Because of this Keith and I had for weeks discussed having a motorboat escort us through these dense patches. Ultimately, we decided to risk it without an escort. As it turned out, we had worried needlessly: what we saw in the aerial photographs were not enormous masses of water hyacinths, but greenish reflections on the water's surface. Which is not to say there were no hyacinths, but there were hardly enough to block our progress.

About 1 p.m. Keith, Ben, and I reached the northeastern extreme of Irish Bend where the antebellum plantation home Oaklawn Manor [29.849652, -91.466709] came into view. Built around 1840 by sugar planter Alexander Porter, the Greek Revival structure survived the Civil War to appear in the 1975 movie *The Drowning Pool*, starring Paul

Newman and Joanne Woodward. Former Louisiana governor Murphy "Mike" Foster Jr. now resided in Oaklawn Manor.

Keith and I noted for some time how few boaters used the Teche for recreation. Whether because of time of year or some other reason, this now changed: that day we saw thirteen pleasure boats on the bayou, one pulling children on an inner tube, others towing children on wakeboards. We also saw someone tooling around on a personal watercraft such as a Jet Ski.

Despite these recreational boaters, the Teche between Baldwin and Franklin impressed me as the most natural, most beautiful stretch of the bayou. As I recorded in my notes that day, "Very quiet when we aren't talking or splashing the oars. Birds whistling and chirping, crickets . . . Sugarcane fields to left or right sometimes, peeking through the tree line following the Teche." (I wrote "splashing the oars," even though we used paddles. As a friend who is a competitive rower once pointed out to me, oars and paddles are not the same thing, nor are rowing and paddling. Rowers sit facing the stern of their boats and use oars for propulsion; paddlers sit facing the bow and use, well, paddles.)

At about 1:15 p.m. we spotted our first alligator since departing Port Barre back in November. It was dead.

We canoed on and on that afternoon, and shortly before 4 p.m. Keith, Ben, and I stopped to speak with a local woman standing on her pier. She and Keith discussed Calumet Cut, also known less poetically as the Wax Lake Outlet, a manmade channel we'd have to cross on our next stage. Keith had mentioned the Cut to me previously and from what I gathered it was not to be taken lightly. His conversation with the woman confirmed my understanding: the current would be fast and treacherous. I did not like the sound of that. Thoughts of Calumet Cut would bother me until we crossed it. I was eager to get it over with.

About 4:05 p.m. we reached Sterling Bridge [29.80242, -91.490204] at Sterling Plantation. Here, on the outskirts of Franklin, we found the modern Sterling Sugars refinery, owned by the sugar-planter Patout family of Patoutville. Ben noticed a makeshift soccer field behind the refinery and conjectured it must provide amusement for seasonal Latin American workers. I later contacted Sterling Sugars and it confirmed Ben's theory. This touches on an ongoing, vibrant

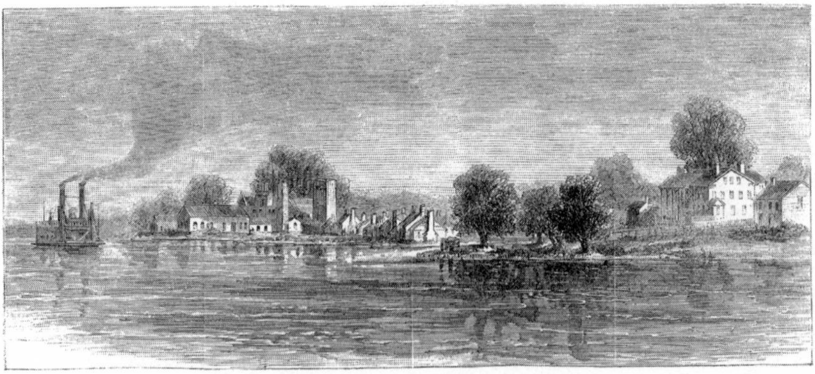

A steamboat on the Teche approaches Arlington, a plantation house near Franklin. The structures to the left of the house appear to be freedmen's quarters and a sugar house complex. Source: *Harper's Weekly* X (December 8, 1866).

cultural change taking place in rural and small-town south Louisiana, including along the Teche corridor: an increasingly large Hispanic population residing in this traditionally French-speaking, now mainly English-speaking region.

We reached Franklin at 4:17 p.m. It was here, on the Teche just as one reaches town, that Confederate captain Semmes scuttled the damaged gunboat *Diana* [approx. 29.798971, -91.496792] after the battles of Fort Bisland and Irish Bend. Set afire by her rebel crew, the artillery-scarred vessel exploded and sank in the bayou near downtown Franklin. In spring 1871 the Army Corps of Engineers removed the *Diana*'s wreckage from the Teche, leaving no visible trace of the gunboat. It could be, however, that numerous small artifacts from the *Diana* still lay at the bottom of the Teche.[23]

Keith, Ben, and I stopped at Franklin [29.791951, -91.499078] just beside the postmodern St. Mary Parish courthouse. Leaving our canoes on the bank, we cut through the courthouse parking lot (glancing at its Confederate memorial, a high-pedestaled statue of a southern infantryman), crossed Main Street (lined with pleasantly archaic globed streetlamps), and stuffed ourselves with food from a corner gas station.

We then embarked again and paddled for another mile down the Teche until we reached our stopping point that day, several acres of land on the bayou owned by my in-laws. Intriguingly, the property sits adjacent to the Greek Revival plantation home called Arlington [29.779823, -91.493144]. And if a mid-nineteenth-century engraving is accurate, it appears Arlington's sugarhouse and slave quarters sat on my in-laws' property. I have walked the tract at length, but found only a fragment or two of nineteenth-century ceramics. I've seen no sign of architectural remains. But my wife's grandfather, recently deceased, once told me that a significant amount of fill had at one time been spread over the property. As a result, any nineteenth-century remains would be buried under at least a few feet of soil.

My in-laws' property sits on the line of withdrawal Taylor's troops used when racing from Fort Bisland to avoid entrapment. My wife's grandmother once showed me a Civil War–era bullet that some metal-detector enthusiasts found on the property. This made sense, because a map in historian Donald S. Frazier's book *Thunder across the Swamp* shows a Union rear-guard action occurring precisely on my in-laws' land. (I have metal-detected there, but so far without luck.)

We debarked at 5:45 p.m. in exhaustion: the 90+°F temperature and smothering humidity had depleted us. (On top of this I suffered sunburns that would pain me for a week.) This was, after all, summer in south Louisiana. And while I still regard that day's stretch of the Teche as my favorite, I also found it the most grueling—if only because we unwisely tackled it in June. That was a mistake. Keith, Ben, and I vowed to wait until fall to undertake the next stage of the Teche. Despite the adverse conditions, we covered 13.6 miles that day, albeit in 8.25 hours—pitifully slow.

Stage 7: Franklin to Calumet

Keith and I set out alone on the seventh stage of our journey down Bayou Teche at 9:40 a.m. on November 3, 2012. We started from where we left off last time, on my in-laws' property just south of Franklin. The morning temperature was 57°F; the afternoon temperature, a balmy 84°F.

We shortly spotted a narrow channel running west from the Teche. It was the Hanson Canal and Lock [29.77261, -91.483286], "[b]uilt in 1907 by the [Albert] Hanson Lumber Company," according to the American Canal Society, "to float log booms from the delta into Bayou Teche." (By "the delta" I assume the ACS meant that of the nearby Lower Atchafalaya River.) In 1922 the Albert Hanson Lumber Company sold the lock to the U.S. government, which entrusted its operation to the Army Corps of Engineers. In 1959 the Corps abandoned the lock, which now sits frozen in the "open" position, a vestige of the Teche region's once prosperous but now extinct logging industry.[24]

By 10 a.m. the sun reflected blindingly off the water. We paddled past suburban homes on the west bank, forest on the east. A glance at an aerial photograph, however, revealed that just beyond that wildwood lay enormous swathes of sugarcane. It was harvest time again—meaning over a year had passed since my canoe team and I began our incremental journey down the Teche. Once more the cane would be cut and ground, its extract boiled and reboiled and processed into raw granulated sugar.

An increasing number of suburban homes on the west bank meant we approached Garden City [29.765279, -91.465889], a village of several dozen houses and a few industrial buildings. A "planned community," it was built in the early twentieth century to house workers of the Albert Hanson Lumber Company. Like Franklin, Garden City makes a short appearance in the classic counterculture movie *Easy Rider*. By "short" I mean three seconds. Those three seconds show an American flag hanging from a whitewashed, wood-framed U.S. Post Office (once used as the lumber company office). Although no longer a public building, the structure still stands, with an American flag hanging from the exact same spot as shown in the movie. This flag went up not too long ago, and if I had to guess I'd say the building's current owner hung it there as an allusion to *Easy Rider*.

Around 10:20 a.m. Keith and I reached Frances Plantation, whose Creole-style "big house," built around 1810, stands on the west bank. Almost directly across from the house Keith and I spotted a rusty iron boiler in the bayou [29.767817, -91.463067], half sunken amid a stand of spindly cord grass. Paddling over to the artifact, we eyeballed it at about four feet in diameter and about twelve feet in length. Although it could have originated in a sugarhouse, we believed it came from a

A steam boiler sits in the water along the east bank of the Teche at Frances Plantation. Is it from a riverboat or a sugar mill? Source: photograph by the author.

steamboat—and that the spot on which it sat encompassed the wreck of a steamboat. We came to this conclusion because several pairs of large iron bolts protruded from the Teche, running in tandem from the boiler toward a few sheets of corroded iron several yards away. A metal cable rested atop these sheets like a dark sunning snake. Using our canoe as a yardstick, Keith and I measured this chain of presumably related artifacts—the boiler, the pairs of bolts, the sheets of metal, the cable—at about 85 feet: the length of a modest-sized steamboat.

I later queried local history enthusiasts about these artifacts and checked inventories of sunken vessels on the bayou. No one knew anything about the site. Thus, the identity of this steamboat—if that's indeed what it is—remains a mystery.

After documenting these artifacts, Keith and I paddled farther down the Teche and soon reached Centerville [29.75987, -91.428401] and right beyond it the community of Verdunville [29.75402, -91.401114], a Creole of Color enclave. Suburban houses stood on both banks. Here we spotted many scattered juglines using cabbage-sized Styrofoam balls as "bobbers." We soon caught up with a motorboater checking these lines. He proudly informed us he'd caught twenty blue catfish so far that day.[25]

After a short distance Keith noticed that the western side of the bayou had become extremely shallow. Sticking his paddle into the hazy water, he touched bottom only about a foot below the surface. Bamboo poles rose from the Teche along this stretch, apparently outlining the shoal for unfamiliar boaters. Aerial photographs indicated that in dry periods this shoal became a lengthy *batture* (which, as mentioned previously, refers to marshy land between the bank of the bayou and the water's edge).

We noticed muddy shoals on both sides of the Teche and, sprouting from these shoals, fingers of aquatic plants twisted underwater in the current like seaweed.

Around noon we passed the mouth of the Verdunville Canal [29.756218, -91.398575], which runs northeast to the West Atchafalaya Basin Protection Levee. Within sight of the canal the Teche dips southward toward Ricohoc and Calumet, clusters of houses taking their names from local sugar plantations.

Calumet Plantation received its name from former postbellum owner Daniel Thompson, who, oddly enough, christened it in homage to the river of the same name flowing through Chicago. Thompson had accrued a fortune operating grain elevators in Chicago before moving to the lower Teche to become a sugar planter. He was among the throng of wealthy northerners who after the Civil War snatched up bankrupt plantations along the bayou and ran them as modern businesses. Unlike some of those northerners, Thompson was genuinely enamored of the Teche Country and chose to live on his plantation until his death.[26]

About 1 p.m. Keith and I steered left at a fork in the bayou [29.705574, -91.378969], paddled our canoe through the West Calumet Floodgate and entered the wide expanse of the Wax Lake Outlet.

Located directly between Ricohoc and Calumet, Wax Lake Outlet [29.701511, -91.372482], sometimes called the Calumet Cut, is a massive manmade channel. Completed in the early 1940s by the Army Corps of Engineers, the Outlet runs over thirteen miles in length and stretches roughly six hundred feet in width. It was designed to spare Morgan City from floods by diverting water from the Atchafalaya swamp to

The west floodgate at Calumet Cut/Wax Lake Outlet. Fortunately for us, both it and the east floodgate happened to be open the day of our trip. Source: photograph by the author.

the nearby Gulf of Mexico. The Corps drove the Outlet across the path of the Teche and, unfortunately, through the remains of the Civil War redoubt known as Fort Bisland [located as best I can determine around 29.700355,-91.373905] and the site of the scuttled Rebel gunboat *Cotton* [approximately 29.699876, -91.373354].

Here the Corps also diverted about a one-mile stretch of the Teche, damming up the bayou's natural bend at Ricohoc/Calumet and replacing it with a straight detour through two modern floodgates. (I have been educated about these structures by retired offshore oilfield fleet owner and local historian F. C. "Butch" Felterman of Patterson; they are not locks, as I initially took to calling them, because they have no chambers in which to raise and lower the water levels; rather, they are floodgates, which close when high water in the Outlet threatens to deluge the Teche.)

The Louisiana Highway Department had previously run Highway 90 through the same narrow neck of high land—the natural Teche Ridge—so that today the Outlet, detour, floodgates, and highway, as well as a preexisting railroad, all nearly come together at a single point. It had to

Crossing the wide Calumet Cut/Wax Lake Outlet (looking northeast toward the Atchafalaya Basin). Source: photograph by the author.

be this way, because except for this ridge the entire area is a flood-prone cypress swamp. This is why the Confederates erected Fort Bisland on this spot: it was an easily defended bottleneck along the Teche.

Wax Lake Outlet seemed vast in comparison to the Teche. Yet we could see on the opposite bank, not too far ahead, the open East Calumet Floodgate and, beyond it, the continuing path of the Teche.

As noted previously, this crossing had concerned me for months because I knew the Outlet's current could be treacherous. In fact, the annual Tour du Teche canoe, kayak, and pirogue race posts "safety boats" in the Outlet in case contestants run into trouble—the only place it does so for the entire competition. Race officials also permit contestants to opt out of crossing the Outlet, without penalty, should they find the passage too dangerous. As a local newspaper reported, "Wax Lake Outlet is the most daunting single obstacle in the Tour. . . . With a sometimes raging current, [it] is nothing to trifle with."[27]

Fortunately, the Outlet was subdued that day, and Keith and I cruised across it as though it didn't even exist. We encountered turbulence only around the two floodgates, where the waters of the Teche and the Outlet converged. There we saw roiling eddies and little whirlpools that quickly swirled into and out of existence. Making strong, deliberate strokes, we paddled through these eruptions and shortly glided into smoother water.

Once through the east floodgate we came to a fork [29.701026, -91.364292]. The right, we knew, was the old Teche channel, now a dead end; the left, however, led farther down the Teche. We steered the latter course and looked for a place to stop for the day. Keith and I had planned to stop at the east floodgate itself—but we unexpectedly found the banks fenced off by the federal and parish governments. So instead we plowed through a mass of reeds and hyacinths to stop at an empty lot next to a house on the east bank. Beyond the lot and house ran Highway 182, which led back to our homes far up the Teche.

It was 1:30 p.m. We had covered a little more than 10.75 miles that day in about 3.75 hours.

While Keith unloaded our supplies, I walked to the house to ask if we could indeed cross the lot to reach the highway. I was told we could do so. We then awaited our ride on the side of the road at Calumet, our backs to the bayou. To our front spread a wide plain covered entirely in stalks of ripe sugarcane. We were standing in the middle of the Fort Bisland battleground [approximately 29.702517, -91.350302]. Although any remains of the fort had been destroyed, most of the fighting had actually occurred about a mile east of the stronghold—right where we stood. If Keith and I had materialized at that exact spot 149 years earlier, we would have been blown away by a crossfire of shell, shot, and minie balls. Despite the Outlet, the highway, and the other modern improvements at Ricohoc/Calumet, the battlefield proper remains, now as then, planted in cane. It would have looked much the same to the Union and Rebel soldiers as it did to us.

How much rusting iron shot and other debris of war lay buried in the field before us, I wondered? How many shells and cannon balls blasted from lumbering field pieces or from the gunboats *Diana* and *Cotton*? And how many scattered bones, perhaps conserved deep in the smothering mud, had we paddled over?

It is well-known among local Civil War buffs that this cane field, like those upstream at Irish Bend, had yielded countless relics: buttons, belt buckles, coins, an array of projectiles. A few months ago, while driving this same stretch of highway, I spotted a construction crew excavating a large hole on the battlefield's edge. I stopped and asked the hard-hatted workers if I might examine the spoil. Sure, go ahead, they said, knocking off for lunch and leaving me alone to poke around the newly exhumed dirt. Surely I would find some artifacts, I thought—a cannonball or a shell fragment, maybe a cache of unspent cartridges. All I found, however, were two decorated ceramic sherds.

Stage 8: Calumet to Berwick

My canoe team and I undertook the final leg of our Teche journey on March 24, 2013. Keith, Preston, and Ben Guidry comprised my crew that day. Keith and I would paddle one canoe, Preston and Ben the other. The sky was cloudless and the temperature at midday a comfortable 70°F, give or take a degree.

Unable to secure permission to start from where we left off previously, we backtracked about a half-mile to the East Calumet Floodgate [29.702319, -91.368753]. This in itself required special permission, because the bayouside there is fenced in, partly by the U.S. Army Corps of Engineers, partly by St. Mary Parish. Figuring it would be easier to deal with the parish than with the federal government, I called my wife's uncle, who retired from the parish, to ask who might give us permission to cross parish property? He connected me with a parish administrator, who gave us the go-ahead to ignore the local "No Trespassing" signs and launch from a cement water control structure next to the federal property.

Parking on Wax Lake Outlet's east levee, we dragged our canoes through a hole in a vandalized hurricane fence—ripped open presumably by trespassers wanting to fish on parish property—and reached the water's edge just below the east floodgate. We put in at 1:07 p.m.

At 1:27 p.m. we passed Zenor Bridge [29.708609, -91.351046], named for the Zenor family of nearby Avalon, Moro, Ingleside, McKay, and Riverside sugar plantations. Here we saw seagulls darting over the water, suggesting our closeness to the open waters of the Gulf of Mexico. Soon we neared the outskirts of Patterson, formerly called Pattersonville

The remains of the *Frenchman*, a 136-foot former Navy mine sweeper that became a commercial "pogie boat." Its resting spot marks the natural mouth of the Teche. Source: photograph by the author.

and, before that, Dutch Settlement. Suburban homes stood on the west bank, while the more flood-prone east bank remained in wilderness. Beyond the woods, however, lay huge swaths of cleared land for growing sugarcane.[28]

Shortly we spotted a low mound of sun-bleached rangia clam shells spilling into the Teche. It appeared to be a midden—the remains of a prehistoric Native American camp or village. One archaeologist, however, has suggested the site is not a midden, but the remnants of a wooden vessel, such as a barge, that happened to sink loaded with shells. Preston, the archaeology graduate, spotted among the shells a few tiny pieces of Native American pottery. This, however, did not necessarily prove the site was originally a midden, because a vessel carrying shells might have pilfered its cargo from a midden, taking away not only the shells, but pottery, bones, and other artifacts. It was, unfortunately, a too common occurrence even into the twentieth century.[29]

Leaving the suspected midden behind, we continued downstream, passing steel barges moored together on the west bank, their black hulls idly scraping together as the wind played over the water. Soon we came to a known midden site, whose rangia shells, like those we saw earlier, were slowly eroding into the waterway. It was the long-abandoned Native American village the Chitimacha call *Qiteet Kuti'ngi na'mu*. This site consists of three earthen mounds and a shell midden. The largest mound rises over nine feet in height and spans nearly a hundred feet in diameter. Preston had excavated at this site during a UL Lafayette field school and told us its history as revealed by the dig. We examined the edge of the mound from our canoes, then continued downstream. (It should be noted that *Qiteet Kuti'ngi na'mu* sits on private property and its artifacts are protected by state law.)[30]

At 2:40 p.m. we arrived at a feature I'd been eager to examine for weeks, noticing its outline in aerial photographs: a large sunken vessel [29.721457, -91.299029]. Parts of its superstructure, blackened by corrosion, protruded here and there from the water, but the hull sat entirely underwater, the lip of its gunwale sitting right below the surface. Chasing leads, months later I tracked down a son of the late John Santos Carinhas, a Portuguese immigrant sea captain and Patterson resident who owned the vessel. According to the son, the vessel was named *The Frenchman*, a 136-foot boat that began its life as a Navy mine sweeper. Under his father, however, it served as a "pogie boat," used for catching a fish species made into pet foods, hog feed, and fertilizers. When *The Frenchman* sank he could not recall, but records indicate it was still afloat as late as 1964.[31]

After examining the remains of *The Frenchman*, we paddled on, unknowingly crossing a major threshold in our journey: we had reached the mouth of Bayou Teche, exited it, and entered a different waterway, the Lower Atchafalaya River. I say unknowingly, though I knew some maps showed the Teche ending there. I had dismissed those maps as incorrect, however, and believed the Teche extended several more miles to the southeast.

It was later that year I met Butch Felterman, who convinced me that the Teche did indeed come to an end at the site of the *Frenchman*'s wreck; and that the waterway below the wreck was the Lower Atchafalaya River.

In retrospect the evidence for this distinction is obvious. Right past the *Frenchman* the distance between the east and west banks flares

Larger ocean-going vessels appeared at Patterson. We were no longer on the Teche, but on the Lower Atchafalaya River. Source: photograph by the author.

from about 365 feet to about 555 feet, and in some spots reaches about 725 feet. Moreover, one can see where, prior to obstruction by the West Atchafalaya Basin Protection Levee, the headwaters of the Lower Atchafalaya—a stretch called Reed's Bayou in the early nineteenth century—swept down from the Atchafalaya swamp and cut across the mouth of the Teche. Finally, present-day maps, as well as the U.S. Army Corps of Engineers and LSU's Cartographic Information Center, assert that the Teche gives way at that spot to the Lower Atchafalaya.

At the time of my final canoe trip, however, I had yet to learn this, and so we paddled on. We soon entered the stretch where in 1863 the USS *Diana* ran a murderous gauntlet of Rebel rifle and artillery fire to end that day a southern prize of war. Today there is nothing to suggest such an incident ever occurred there.[32]

More seagulls wafted overhead and shortly we glided under Bridge Road to reach Patterson [29.693363, -91.300651]. The town, I should note, was named for a certain "Captain Patterson" who commanded vessels up and down the bayou. Several historical documents refer to the captain, but do not mention his first name—almost as if to suggest that everyone knew the captain, so why bother? Some modern sources identify him as John Patterson, but I can find no references to John

Patterson in pertinent historical documents. I did find a few references to Elam Patterson, a known Pattersonville resident who ferried sugar from St. Mary Parish to New Orleans aboard keelboats and steamers, and who died around 1851. It is my belief that Elam was the Captain Patterson for whom the town was named—but this is speculation.[33]

Moored on the river at Patterson we saw offshore supply vessels, survey boats, utility boats, crew boats, and pushboats that dwarfed our canoes. Passing houses and businesses on the west bank, we eventually came to Butch Felterman's private museum, dubbed Fishermen's Light [29.692334, -91.301260]. Dedicated to Patterson's nautical history, it highlights the local shrimping industry that operated from 1937 to 1988. Mr. Felterman's museum is easy to spot: he built the multi-story structure to look like an oceanside lighthouse. Moreover, a roughly eight-foot colonial-era cannon of pockmarked iron sits in the yard and is aimed menacingly at the waterway. (The cannon is not a local artifact, but came from the wreck of *El Nuevo Constante*, a Spanish ship that foundered off the Louisiana coast in 1766.)[34]

At 3:41 p.m. we began to encounter brisk winds and choppy waves. This seemed unusual for a river. The Lower Atchafalaya, however, was more like a lake, not only because of its expansive width, but because water control structures sealed off its extremes. This effectively converted the stream into a lake.

At 4:15 p.m. Keith and I saw a large alligator. And it saw us. It slipped from its perch of driftwood grounded near the wooded east bank and swam rapidly in our direction. I had not expected such an apparent show of aggression. The gator submerged as it slid like a torpedo toward us and I held my breath, waiting for its spiked back to scrape against our canoe's bottom. But I heard and felt nothing. It was gone.

A little after 5 p.m. we spotted a bald eagle circling overhead. It occasionally plummeted to snatch fish from the Lower Atchafalaya. This stretch of waterway—in fact, this entire section of St. Mary Parish—is home to many bald eagles. I have seen bald eagles along nearby Bayou Sale (pronounced "Sally" by locals, though no doubt originally pronounced "Salé" in the French manner, meaning "Salty"). There, along the West Bayou Sale Levee, I saw a bald eagle's nest—a massive aerie in the top of a towering oak or cypress.

An explosion of water, rising several feet next to the canoe like a small geyser, interrupted our birdwatching. "What the hell was that!" Something large had breached the surface, either another alligator or an enormous fish. It spooked us. We'd never seen anything like it before on our trips.

My team and I soon reached Bayou Vista [29.690981, -91.274109], a community I had never before visited, and in my ignorance I imagined it a shadowy backwoods. To my surprise, however, luxury boats and modern dwellings sat along the manicured west bank.

While paddling through Bayou Vista, Keith and I noticed Preston and Ben falling farther and farther behind us. We kept glancing over our shoulders at them, but with the setting sun in our eyes all we could make out was the glint of their aluminum canoe. We assumed they had grown tired, and felt puffed up that we, the two older guys, were so far ahead, even on such a blustery afternoon in rough water.

We had no idea, however, that Preston and Ben had flipped their canoe. They had gone into the water, along with Ben's utility ax, a celebratory bottle of champagne, and several hundred dollars' worth of electronics. Fortunately, a fisherman had witnessed the spill and quickly pulled Preston and Ben from the water.

Keith and I learned about the accident only when Preston and Ben called us by cell phone. Should we turn back? No, Keith decided, we had to push forward, because our vehicle sat only a short distance ahead at Berwick. In the meantime, Keith made arrangements for a friend to pick up Preston and Ben along Highway 182 and take them to shelter back in Patterson.

Keith and I paddled on largely in silence, thinking about poor Preston and Ben huddled on the side of the road in wet clothing. Soon we past what appeared to be another shell midden, this time on the west bank—but there was no time to examine it. I jotted down the coordinates and we kept going.

As we approached the northern limits of modern Berwick, we reached the former site of Rentrop's tavern and ferry landing. It was here, on the western edge of the great Atchafalaya swamp, that early-nineteenth-century travelers relied on a state-sanctioned ferry monopoly. Granted to Henry Rentrop (sometimes spelled Renthrop) and his heirs, this ferry shuttled passengers and livestock, according to the 1811 state license, "across the lakes Plat [Flat Lake], Palourde, de Jonc and Verret, from the lower part of the Bayou Teche [now the Lower Atchafalaya] to the Bayou Verret, in the county of Lafourche." Cathcart, the U.S. Navy agent and surveyor, lodged at Rentrop's tavern in 1819 and observed in his journal:

> Mr. Renthrop & his son are tailors, natives of Westphalia, came to Philadelphia some years ago & have traveled through many places in the United States since, & about nine years ago settled upon this spot. They keep a tolerable good table for this part of the world. Their beds are clean, provisions wholesome. Liquors [are] whiskey, tafia & bad claret. They are obliging but wholly illiterate. Their farm is not very extensive, but their garden is productive. They raise poultry & hogs in abundance & some fine cattle, & this is the first place we had milk with our coffee since we left New Orleans. Fresh butter is entirely out of the question, & salt cannot be procured except in the City. Hogs lard is made its substitute in all culinary purposes. . . . The flats (so call'd) used at this Ferry, are form'd of two large canoes, on which is a platform for horses. The price of carriage for a man & horse is 12 dollars, & for black cattle 1.50 cs per head. They cross the Lake [Lake Palourde] to the [Attakapas] canal which runs into Lake Verret from Lafourche, a distance of 30 miles, & from thence passengers proceed to Donaldsonville, & take passage in steamboats, that pass either up or down the Mississippi. . . .[35]

At 5:35 p.m. we sighted the West Atchafalaya Basin Protection Levee looming in front of us; two minutes later we spotted the Berwick Lock [29.719312, -91.224536]. On reaching the cement-and-iron structure, I used my cellphone to call the lockmaster. Shortly a man leaned over the railing above us, shouted a greeting, and disappeared to open the watertight inner gate. As soon as the lock opened, we paddled inside. Large enough to accommodate those sizable vessels back at Patterson, the lock made our canoe seem ridiculously small. The lockmaster cracked the outer gate to let in about a foot of water, then fully opened the outer gate. Keith and I then paddled out into a much wider section of the Lower Atchafalaya River, which here turned south toward the Gulf. Stretching about 1,700 feet across, the

The terminus of our 135-mile journey down the Teche and part of the Lower Atchafalaya River: the 155-year-old Southwest Reef Lighthouse at Berwick. Source: photograph by the author

river's banks were lined with industrial docks, barges, jack-up boats, pushboats, and other large vessels.

Steering downstream, we canoed the rough waters of the wide, turbulent river and soon passed underneath the eighty-year-old steel-truss Long-Allen Bridge [29.696688, -91.214381] and the parallel, more modern E. J. "Lionel" Grizzaffi Bridge. Beyond the bridges we pulled up to Berwick's 155-year-old iron-plated Southwest Reef Lighthouse [29.694273, -91.216313], which stood offshore in Atchafalaya Bay until moved to this site in 1990 for preservation.

It was about 6 p.m. and we had covered 15.5 miles in approximately five hours.

As it turned out, not only had the two Guidry brothers lost their electronics and other belongings when they overturned, but Preston had suffered a cracked rib during his rescue. He ended the day in a hospital emergency room in Lafayette and did not arrive home until after 1 a.m. the next morning.

Regardless, we had done it—though of five paddlers Keith and I were the only ones to complete all eight stages of the journey. Still, we couldn't have made it without the others' help. Over the past year and a half we had spent approximately 47.5 hours—30 minutes short of two full days—paddling from Port Barre to Berwick. It had taken us 44.25 hours, however, to reach the Teche's mouth back at Patterson.

On our journey we learned that the Teche is underutilized—perhaps a blessing, particularly for those who already use the bayou for boating, fishing, and other forms of recreation despite its "impaired" status. We also learned that, even with zealous clean-up campaigns, litter still taints the banks and blots the country lanes along the waterway. More positively, however, we discovered some stretches of the Teche to be pristine, or nearly so. Additionally, we realized that the Teche is one enormous ecosystem: that life begins, thrives, and dies in its waters and in the very ooze along its banks, from microscopic organisms to fish, turtles, and alligators. I am hopeful that ongoing efforts to quash litter and pollution will improve the Teche significantly in coming years. Apathy and ignorance will as usual collude to block progress, but I believe most people, regardless of cultural background, economic status, educational attainment, or political affiliation, want a clean Bayou Teche.

Acknowledgments

At the University of Louisiana at Lafayette I wish to thank Carl A. Brasseaux of the Center for Louisiana Studies (retired) for answering my occasional questions; Barry Jean Ancelet of the Department of Modern Languages (retired) for assistance with my French translations; Bruce Turner of Special Collections and Jean S. Kiesel of the Louisiana Room, both in Edith Garland Dupré Library, for responding to my queries; Whitney P. Broussard III of the Institute for Coastal Ecology and Engineering for help with environmental data; Garrie P. Landry of the Department of Biology for discussing plant and animal life on the bayou; Michael S. Martin and James D. Wilson of the Center for Louisiana Studies for responding to my random queries; and Charles E. Richard of the Department of English for advising me about films pertaining to the Teche.

In the Anthropology program at UL Lafayette I thank Mark Rees and David Palmer for discussing Teche archaeology; Jim Delahoussaye for advising me about plant and animal life on the bayou; C. Ray Brassieur for sharing information about Promised Land cemetery and other subjects; and Jon L. Gibson (retired) for providing documents about the Chitimacha tribe.

From other institutions, I extend my appreciation to Donald W. Davis, Louisiana Sea Grant Scholar, Louisiana State University, for explaining the geography of the Teche; Elaine Smyth of LSU Libraries (retired) for responding to my archival inquiries; John M. Anderson of LSU's Cartographic Information Center and Ricky D. Boyett of the U.S. Army Corps of Engineers for help with map queries; Donald S. Frazier of the Department of History, McMurry University, for proofing my Civil War chapter; Louisiana State Archivist Florent Hardy Jr. for allowing me access to the journal of his ancestor, snag boat superintendent

Jean Jules Hardy; Rogelio Saenz of the University of Texas at San Antonio and Julia M. Bernard of McNeese State University for assisting with Spanish translations; and Jack B. Martin of the Departments of English and Linguistics, College of William and Mary, and Daniel W. Hieber of Rosetta Stone for answering my questions about Native American linguistics and for critiquing my theories about the possible origins of the word *Teche*.

I also wish to thank R. Christopher Goodwin and Nathanael Heller of R. Christopher Goodwin & Associates and Charles E. Pearson of Coastal Environments Inc. for sharing information about Teche archaeology; Elaine Clément of the Acadian Memorial and Department of Tourism, St. Martinville, and independent genealogist Jane Guillory Bulliard for proofing my overview of early St. Martinville history; independent researcher George Bentley for sharing his knowledge of Teche history; independent researcher David Lanclos for translating French documents, for proofing my French translations, and for critiquing my maps; David Cheramie of the Bayou Vermilion District and educator Faustine Hillard for critiquing my French translations; genealogists Stanley LeBlanc, Judy Riffel, and Winston De Ville for answering questions, pursuing elusive colonial documents, and, in Stanley's case, proofing parts of the manuscript and critiquing my maps; Margie Luke of the St. Mary Chapter Louisiana Landmarks Society for sharing an extremely important colonial document and for discussing lower Teche history; Conni Castille, Kristen Kordecki, Dane Thibodeaux, Patti Holland, and Mena LeBlanc with the TECHE Project, Ken Grissom of the Tour du Teche, and Blake Couvillion of Cajuns for Bayou Teche for discussing recent efforts to clean and conserve the Teche; Dana L. White of the Environmental and Regulatory Services Division, Austin Water Utility, City of Austin, Texas, for helping me to understand waterborne pollutants; Glen Pitre of Côte Blanche Productions for sharing information about Teche film depictions; archaeologist Donny Bourgeois for sharing his knowledge of archival maps and other subjects; Charles Larroque of the Council for the Development of French in Louisiana and Ronald Gaspard for information about Promised Land cemetery; Roland R. Stansbury of the Young-Sanders Center for the Study of the War between the States in Louisiana for providing Civil War maps and pointing me toward sources about the Teche campaigns; Reving Broussard for permitting me access to the Lutzenberger Foundry; Kim Walden, Roger Stouff, and Alton LeBlanc of the Sovereign Nation of the Chitimacha for sharing tribal history; Benny Graeff for sending the lyrics to the song “Port Barre”; Mary Wyche Estes and Guy Estes for allowing me access to the watermill ruins at Belmont Plantation; F. C. “Butch” Felterman for sharing information about the lower Teche and Lower Atchafalaya River; Kurt Crosby of Crosby Tugs for advising me about modern Teche pushboats; the late Mayor Al Broussard of Loreauville for answering questions about local Indian mounds and other topics, and for permitting my canoe team to access city property; J. P. Hebert, Donovan Garcia, and Dave Lowery for answering questions about the lower Teche; Nancy Lees for sharing historic newspaper articles; Henry C. “Bo” LaGrange, chief administrative officer of St. Mary Parish Government, for helping my canoe team access the bayou at Calumet; Melissa Bonin for advising me about the Teche in visual art; Willy Bernard for sharing his findings about the Flood of 1927; Jennifer Miller and the Miller family of Irish Bend for permitting me to copy the ship’s logs of their ancestor, steamboat captain William H. M. Lynch; Thomas Kramer, Chris Freeman, and Stephen Stirling for discussing St. Mary Parish history; Smitty Landry for sharing tales of the Teche from his youth; Donald Hahn and the Hahn family of Franklin for helping my canoe team access the lower Teche; Jody Carinhas for providing information about his father Captain John Santos Carinhas and *The Frenchman*; Warren A. Perrin for covering the cost of acquiring the 1942 film *Cajuns of the Teche*; and the Iberia Parish Library for its interlibrary loan program and other invaluable services.

I offer special thanks to Keith Guidry, Preston Guidry, Ben Guidry, and Jacques Doucet for canoeing the Teche with me (in canoes kindly provided by the Guidrys); and Donald Arceneaux for proofing my manuscript, sharing his research on Acadian settlements along the Teche, paddling a stretch of the Teche with me, and discussing Teche Country history for countless hours.

At University Press of Mississippi I thank Craig Gill, Katie Keene, and Steve Yates, among others, with whom I have enjoyed working on this and previous books.

I am extremely grateful to McIlhenny Company and Avery Island Inc., whose generosity as my employers made it possible for me to undertake this project.

And finally I thank my children Colette and Alexandre for inspiration; and my wife Amy Lançon Bernard for patience and encouragement over the seven years I worked on this project.

Notes

Introduction

1. U.S. National Park Service, "Mississippi River Facts," National Park Service website, www.nps.gov/miss/riverfacts.htm, accessed November 27, 2012.

Teche discharge figure derives from average discharges for the years 1997 to 2013, as measured by the U.S. Geological Survey at Adeline Bridge near Jeanerette, Louisiana (including years for which there is incomplete data). U.S. Geological Survey, National Water Information System: Web Interface, Surface Water Data for USA: USGS Annual Statistics, waterdata.usgs.gov/nwis/annual?, accessed February 5, 2015.

Measurements for the widths of the Mississippi River and Bayou Teche were derived from Google Maps (maps.google.com) by applying its distance measurement tool to aerial photographs.

Unless otherwise indicated I have modernized and standardized the spelling and punctuation of all primary-source quotations in this book.

2. William Darby, *Geographical Description of the State of Louisiana* (New York: James Olmstead, 1817), 142; William C. C. Claiborne, County of Attakapas, LA, to President Thomas Jefferson, Washington, D.C., July 25, 1806, in Clarence Edwin Carter, comp. and ed., *The Territorial Papers of the United States, Vol. IX: The Territory of Orleans, 1803-1812* (Washington: U.S. Government Printing Office, 1940), 678; Willard McKinstry, "The Bayou Teche," *Selections of Editorial Miscellanies and Letters, Published in the Fredonia Censor at Various Times between 1842 and 1894* (Fredonia, NY: Censor Printing Office, 1894), 173; Harnett T. Kane, *The Bayous of Louisiana* (New York: William Morrow, 1943), 226.

3. *Cajuns of the Teche* (Columbia Pictures, 1942), film reel in the National Archives and Records Administration, NAIL Control Number NWDNM(m)-306.7469, digital copy in possession of the author, available for viewing at Shane K. Bernard, "'Cajuns of the Teche': Bad History, Wartime Propaganda, or Both?" *Bayou Teche Dispatches*, blog, May 9, 2013, bayoutechedispatches.blogspot.com/2013/05/cajuns-of-teche-bad-history-wartime.html; Euphemia E. Fleurot, "Souvenir du Teche Polka," sheet music (New York: William Hall and Son, 1856); John Phillip Sousa, "The Belle of Bayou Teche," lyrics by O. E. Lynne, sheet music (Cincinnati: John Church, 1911).

4. Columbus Frugé, "Valse de Bayou Teche," in Raymond E. François, ed., *Yé Yaille, Chère!: Traditional Cajun Dance Music* (Lafayette, LA: Thunderstone Press, 1990), 39–40; Nathan Abshire, "Valse de Bayou Teche," in François, *Yé Yaille, Chère!*, 41–42.

5. Henry Wadsworth Longfellow, *Evangeline and Selected Tales and Poems* (New York: Signet Classics/New American Library, 1964), 82; Charles C. Calhoun, *Longfellow: A Rediscovered Life* (Boston: Beacon Press, 2005), 25; Carol A. Kolmerten, "Introduction," in Alice Ilgenfritz Jones and Ella Merchant, *Unveiling a Parallel: A Romance* (Syracuse, NY: Syracuse University Press, 1991), xiv; James Lee Burke, *Crusader's Cross* (New York: Pocket Books/Simon & Schuster, 2005), 115.

The forgery of an 1822 memoir of a 1795 journey up the Teche is examined in J. John Perret, "Strange True Stories of Louisiana: History or Hoax," *Southern Studies* 16 (Spring 1977): 41–53; and Shane K. Bernard, "A 1795 Journey up the Teche: Fact, Fiction, or Literary Hoax?" *Bayou Teche Dispatches*, blog, August 30, 2011, bayoutechedispatches.blogspot.com/2011/08/in-1889-famed-louisiana-author-george.html.

6. Concession to Jean-Baptiste Grevemberg, cover page dated July 16, 1765, ADS, Grevemberg Family Documents, Americana Collection, National Society of the Daughters of the American Revolution, Washington, DC, digital copy in possession of the Grevemberg House Museum, St. Mary Chapter Louisiana Landmarks Society, Franklin, LA.

See also "Memorial of Gravemberg [*sic*]," July 16, 1765, and reply by Aubray [*sic*] and Faucault [*sic*], n.d., Pintado Papers [Vicente Sebastián Pintado Papers], Opelousas District, Book 1, Pt. 2, 159 [typewritten English translation], microfilm in the St. Martin Parish Library, St. Martinville, LA.

7. "The Bayou Teche—Who Was It Named After," *Opelousas Courier*, October 8, 1859, reprinted in *Attakapas Gazette* XV (Summer 1980), 52; Norman McF. Walker, "The Geographical Nomenclature of Louisiana," *Magazine of America* X (September 1883): 215–16; Melissa Braud, Benjamin Maygarden, and Rhonda Smith, *Cultural Resources Survey of a Borrow Area for the West Atchafalaya Basin Protection Levee Item W-123, St. Mary Parish, Louisiana* (New Orleans: Earth Search, September 1998), 12.

8. *The Bayou and Its People* (New Orleans: Southern Pacific Lines, 1922), 6; Hilda Roberts, comp., "Louisiana Superstitions," *Journal of American Folk-lore* 40 (April-June 1927): 144.

9. In his unpublished "A Sketch of the Chitimacha Language," Swanton rendered the Chitimacha's name for Bayou Teche as *ukc tcat* (Snake Bayou), which is an alternate spelling of *qukx caqaad*. John R. Swanton, "A Sketch of the Chitimacha Language," [1920], TD, 49, copy in the Jean Lafitte National Historical Park Collection, Louisiana and Special Collections Department, Earl K. Long Library, University of New Orleans, New Orleans.

10. The word for *snake* in Chitimacha is *qukx* or *ukc*; in Attakapas, *ōtsi* and *natkoi*; in Choctaw, *sinti*; in Houma, *sănté*; in Koasati (Coushatta), *cintó*; and in Tunica, *nara*.

John R. Swanton, *A Structural and Lexical Comparison of the Tunica, Chitimacha, and Atakapa Languages*, Bureau of American Ethnology, Smithsonian Institution, Bulletin 68 (Washington: U.S. Government Printing Office, 1919), 45, 55; John R. Swanton, *Indian Tribes of the Lower Mississippi Valley and Adjacent Coast of the Gulf of Mexico*, Bureau of American Ethnology, Smithsonian Institution, Bulletin 43 (Washington: U.S. Government Printing Office, 1911): 28–29, 352; Cyrus Byington, *A Dictionary of the Choctaw Language*, ed. John R. Swanton and Henry S. Halbert, Bureau of American Ethnology, Smithsonian Institution, Bulletin 46 (Washington: U.S. Government Printing Office, 1915), s.v. "snake";

Geoffrey D. Kimball, *Koasati Dictionary* (Lincoln: University of Nebraska Press, 1994), s.v. "snake"; Jack Martin, linguist, College of William and Mary, Williamsburg, VA, e-mail to the author, January 26, 2013; Daniel W. Hieber, linguist, Rosetta Stone, Harrisonburg, VA, e-mail to the author, January 29, 2013.

11. Mary Avery McIlhenny Bradford (Mrs. Sidney Bradford), Chitimacha basketry notebook, n.d., Museum of Natural Science, Louisiana State University, Baton Rouge.

For more on my "worm" theory, see my blog article, "A Snake, a Worm, and a Dead End: In Search of the Meaning of 'Teche,'" *Bayou Teche Dispatches*, January 30, 2012, bayoutechedispatches.blogspot.com/2013/01/a-snake-worm-and-dead-end-in-search-of.html.

12. "Carta de Don Damian Manzanet a Don Carlos de Siguenza sobre el Descubrimiento de la Bahía del Espiritu Santo [1690]," *Quarterly of the Texas State Historical Association* II (April 1899): 258; Lilia M. Casís, trans., "Letter of Don Damian Manzanet to Don Carlos de Siguenza Relative to the Discovery of the Bay of Espiritu Santo [1690]," *Quarterly of the Texas State Historical Association* II (April 1899): 286.

Casís's English translation of Manzanet's letter contains a typographical error perpetuated by subsequent authors. Namely, it states that Native Americans shouted "thechas! techas!"—the second word beginning with "t" not "th." The accompanying Spanish-language transcript, however, indicates that the phrase should have read "thechas! thechas!"

13. John R. Swanton, *Source Material on the History and Ethnology of the Caddo Indians* (Norman: University of Oklahoma Press, 1996), 4.

14. José Antonio Pichardo, *Pichardo's Treatise on the Limits of Louisiana and Texas, Vol. III*, Charles Wilson Hackett, trans., ed., and anno. (Austin: University of Texas Press, 1941), 46.

15. Unfortunately, it is unclear which of the two rivers, the Sabine or the Mermentau, the Spanish regarded as the Rio Mexicano. Indeed, the Spanish themselves may not have agreed on the designation. It is even possible that some Spanish used Rio Mexicano in reference to another candidate river, the Calcasieu.

The Sabine, however, is associated with the Rio Mexicano in Elliott Coues, *The Expeditions of Zebulon Montgomery Pike, Vol. 2: Arkansas Journey-Mexican Tour* (New York: Francis P. Harper, 1895), 702 n7. Writing even earlier (1812), Pichardo indirectly associates the Sabine with the Rio Mexicano by stating that the Mexicano is the same as the Rio de los Adaes, which others in turn have identified as the Sabine. See Pichardo, *Pichardo's Treatise on the Limits of Louisiana and Texas, Vol. IV*, Charles Wilson Hackett, trans., ed., and anno. (Austin: University of Texas Press, 1946), 225; compare with Conrad Malte-Brun, *La Geografía Universal, Tome II* (Madrid: Librería, 1853), 344.

The Rio Mexicano is associated with the Mermentau River in Jedidiah Morse, *A Compendious and Complete System of Modern Geography* (Boston: Thomas and Andrews, 1814), 260, 269.

In addition, the Rio Mexicano was mentioned as a western Louisiana river as early as 1778. See the anonymously penned *The Present State of the West-Indies* (London: R.

Baldwin, 1778), 15, which referred to "the Rio Mexicano, which divides Louisiana on the west from New Mexico. . . ."

16. For more on the origin of the word *gumbo*, see my blog articles "Gumbo in 1764?" and "On That Word 'Gumbo': Okra, Sassafras, and Baudry's Reports from 1802-1803," *Bayou Teche Dispatches*, bayoutechedispatches.blogspot.com, October 3, 2011, and October 19, 2011.

17. Raymond Fogelson and William Sturtevant, eds., *Handbook of North American Indians, Vol. 14: Southeast* (Washington: Smithsonian Institution, 2004), 175; Byington, *Dictionary of the Choctaw Language*, s.v. "bok."

18. Shane K. Bernard, *The Cajuns: Americanization of a People* (Jackson: University Press of Mississippi, 2003), xix; Shane K. Bernard, "Cajuns," *KnowLA Encyclopedia of Louisiana*, www.knowla.org/entry/626/&view=article, accessed December 22, 2012.

19. Bernard, *The Cajuns*, xxiv; Shane K. Bernard, "Creoles," *KnowLA Encyclopedia of Louisiana*, www.knowla.org/entry/627/&view=article, accessed December 22, 2012.

20. Glenn R. Conrad, "A Road for Attakapas," *Attakapas Gazette* XV (Spring 1980): 30; Glenn R. Conrad, *Land Records of the Attakapas District, Vol. I, The Attakapas Domesday Book: Land Grants, Claims and Confirmations in the Attakapas District, 1764-1826* (Lafayette: Center for Louisiana Studies, University of Southwestern Louisiana, 1990), 4, see also footnote 2 on same page; "Survey of Bayou Teche from St. Martinville to Port Barre, Louisiana," *Annual Reports of the War Department for the Fiscal Year Ended June 30, 1897, Report of the Chief of Engineers, Part 2* (Washington: U.S. Government Printing Office, 1897), 1786; "Report of Major W. H. Heuer, Corps of Engineers, Office in Charge, for the Fiscal Year Ending June 30, 1886, with Other Documents Relating to the Works," *Annual Report of the Chief of Engineers, United States Army, to the Secretary of War, for the Year 1886, Pt. II* (Washington: U.S. Government Printing Office, 1886), 1260.

21. All recent population data for communities along the Teche derives from the U.S. Census Bureau, 2010 Population Finder, www.census.gov/popfinder/, accessed December 30, 2012, or U.S. Census Bureau, American FactFinder, www.factfinder.census.gov/, accessed February 5, 2015.

22. "Bayou Teche Locks," *New Iberia Enterprise and Independent Observer*, June 14, 1913, 2; "Keystone Locks," (St. Martinville, LA) *Weekly Messenger*, July 19, 1913, 3; "Southwest Louisiana's Biggest Project Nearing Completion," (Opelousas, LA) *St. Landry Clarion*, September 4, 1920, 1.

23. David C. Edmonds, *Yankee Autumn in Acadiana: A Narrative of the Great Texas Overland Expedition through Southwestern Louisiana, October-December 1863* (Lafayette, LA: Acadiana Press, 1979), 216–17.

24. "Report of Major W. H. Heuer, Corps of Engineers, Office in Charge, for the Fiscal Year Ending June 30, 1886, with Other Documents Relating to the Works," *Annual Report of the Chief of Engineers, United States Army, to the Secretary of War, for the Year 1886, Pt. II* (Washington: U.S. Government Printing Office, 1886), 1248; "Public Sale," for land "on the north side of the Bayou Grand Louis or Teche, or Carron," *Opelousas Journal*, April 1, 1871, 2; Undated map labeled "Bayau [*sic*] Catereau or Teche," digital copy in the possession of

Mena LeBlanc, historian for the TECHE Project, Arnaudville, LA, copy in possession of the author.

25. Donald Arceneaux, Moscow, Idaho, e-mail to the author, New Iberia, LA, 30 January 2013.

26. "The Sugar Plantations of Louisiana," (Franklin, LA) *Planters' Banner*, 26 January 1870, 2. See also Paul F. Stahls Jr. and Leonard Hingle, *Plantation Homes of the Teche Country* (Gretna, LA: Pelican Publishing, 1979).

27. Carl A. Brasseaux, *In Search of Evangeline: Birth and Evolution of the Evangeline Myth* (Thibodaux, LA: Blue Heron Press, 1988), 23, 45.

28. *Louisiana, 1925-1926* (Baton Rouge: Louisiana Department of Agriculture and Immigration, 1926), 172; (St. Martinville, LA) *Weekly Messenger*, September 2, 1899, 2.

29. Karma Champagne, "Using Robot Power: Foundry, 158 Years Old, Stays Abreast of Technology," (New Iberia, LA) *Daily Iberian*, special "Centers of Industry" section, February 25, 2010, 1, 10.

30. Thomas Frère Kramer, *A Family Montage: Being the Families Kramer, Frère, Foster, Marsh, Gates, and Allied Families* (Lafayette: Center for Louisiana Studies, University of Louisiana at Lafayette, 2002), 197; Thomas Frère Kramer, Franklin, LA, telephone interview by the author, February 3, 2015.

31. Carl A. Brasseaux, Glenn R. Conrad, R. Warren Robison, *The Courthouses of Louisiana*, 2nd ed. (Lafayette: Center for Louisiana Studies, University of Southwestern Louisiana, 1997), 164.

32. Walter Prichard, ed., "Some Interesting Glimpses of Louisiana a Century Ago," *Louisiana Historical Quarterly* XXIV (January 1941): 40.

33. John M. Anderson, Map Librarian and Director, Cartographic Information Center, Department of Geography and Anthropology, Louisiana State University, Baton Rouge, e-mail to the author, January 22, 2013; Ricky Boyett, Public Affairs Office, New Orleans District, U.S. Army Corps of Engineers, New Orleans, e-mail to the author, January 22, 2013.

Felterman and the Patterson proclamation are quoted in Harlan Kirgan, "9.4M Patterson Water Plant Approved," StMaryNow.com, March 12, 2014, www.daily-review .com/local/94m-patterson-water-plant-approved, accessed March 17, 2014.

Chapter 1

1. O. Abington, H. Bullamore, and D. Johnson, *Louisiana: A Geography* [Lafayette: Department of Geography, University of Southwestern Louisiana, 1989], 11, 15, 22–23, 24; Darwin Spearing, *Roadside Geology of Louisiana*, 2nd ed. (Missoula, MT: Mountain Press, 2007), 38, 67; R. Christopher Goodwin et al., *Overview, Inventory, and Assessment of Cultural Resources in the Louisiana Coastal Zone* (New Orleans: R. Christopher Goodwin and Associates, 1991), 38, 53; Mark A. Rees, Lafayette, LA, e-mail to the author, March 1, 2015.

2. Joseph S. Meyer et al., *Cultural Resources Survey of Four Disposal Areas along the Vermilion River, Lafayette Parish, Louisiana* (Fort Walton Beach, FL: Prentice Thomas and

Associates, 1995), 19–20; Shane K. Bernard, *Tabasco: An Illustrated History* (Avery Island, LA: McIlhenny Company, 2007), 222; Mark A. Rees, "Plaquemine Mounds of the Western Atchafalaya Basin," *Plaquemine Archaeology*, eds. Mark A. Rees and Patrick C. Livingood (Tuscaloosa: University of Alabama Press, 2007), 70, 79 (Table 4.1); Mark A. Rees, Lafayette, LA, e-mail to the author, April 7, 2013.

3. Swanton, *Indian Tribes of the Lower Mississippi Valley*, 360–61, 361 (fn. a), 363–64; Alfred E. Lemmon, John T. Magill, and Jason R. Wiese, eds., *Charting Louisiana: Five Hundred Years of Maps* (New Orleans: Historic New Orleans Collection, 2003), 58 (map 18), 61 (map 21); Richard A. Weinstein, David H. Kelley, and Joe W. Saunders, eds., *The Louisiana and Arkansas Expeditions of Clarence Bloomfield Moore* (Tuscaloosa: University of Alabama Press, 2004), 138–39.

4. Rees, "Plaquemine Mounds," 84–87; Rees, e-mails to the author, April 7, 2013, March 1, 2015.

Swanton observed: "In a political sense [the word *Atakapa* (*Attakapas*)] came to designate a district embracing the present parishes of St. Mary, Iberia, St. Martin, Lafayette, and Vermilion. From this it might seem as if the Atakapa had once occupied the entire region, but according to the best evidence St. Mary and the eastern parts of Iberia and St. Martin were in Chitimacha territory." Swanton, *Indian Tribes of the Lower Mississippi Valley*, 360.

5. Swanton, *Indian Tribes of the Lower Mississippi Valley*, 343–44.

6. Thomas Hutchins, *An Historical Narrative and Topographical Description of Louisiana and West-Florida* (Philadelphia: Robert Aitken, 1784; repr., Gainesville: University of Florida Press, 1968), 46; Swanton, *Indian Tribes of the Lower Mississippi Valley*, 343; Mathé Allain, "Bouligny's Account of the Founding of New Iberia [part one]," *Attakapas Gazette* XIV (Summer 1979): 83.

7. See Roger Stouff and Dayna Bowker Lee at 15 min. 25 sec. in *Native Waters: A Chitimacha Recollection* (Baton Rouge: Louisiana Public Broadcasting, 2011), DVD; Roger Stouff, n.p., e-mail to the author, April 3, 2011, August 11, 2011.

8. [Vincent H. Cassidy], "The De Soto Expedition," *Attakapas Gazette* II (June 1967): 17.

9. "The Narrative of the Expedition of Hernando de Soto by the Gentleman of Elvas," in *Spanish Explorers in the Southern United States, 1528–1543*, Theodore H. Lewis, ed. (New York: Charles Scribner's Sons, 1907), 261. See also [Cassidy], "The De Soto Expedition," 18–19.

10. Herbert Eugene Bolton, *Texas in the Middle Eighteenth Century: Studies in Spanish Colonial History and Administration* (Berkeley: University of California Press, 1915), 360, 361; Abbé [Pierre?] Didier, [n.p.], to Agustin Ahumada y Villalon, Marques de las Amarillas, Viceroy of New Spain, July 19, 1756, Herbert F. Bolton Papers, Bancroft Library, University of California, Berkeley [typewritten transcript, original document now missing in the Archivo General de la Nación, Mexico City, Mexico]; Jacinto de Barrios y Jauregui, Governor of Texas, to Agustin Ahumada y Villalon, Marques de las Amarillas, Viceroy of New Spain, July 22, 1756, cover letter accompanying Didier's aforementioned correspondence, Bolton Papers [typewritten transcript, original document now missing in the Archivo General de la Nación, Mexico City, Mexico].

Writing in the late nineteenth century, Francis DuBose Richardson and Alcée Fortier both wrongly conflated the Attakapas and the Chitimacha. Francis DuBose Richardson, "The Teche Country Fifty Years Ago," *Southern Bivouac* (January 1886), reprint, Glenn R. Conrad, ed., *Attakapas Gazette* VI (December 1971): 128; Alcée Fortier, "The Acadians of Louisiana and Their Dialect," *Publications of the Modern Language Association* VI (1891), reprinted as "The Acadians of Louisiana," *Attakapas Gazette* XXVI (Fall 1991): 136.

11. Hutchins, *An Historical Narrative*, xix, xl–xli, xliii, 46–47; Thomas Hutchins, "Courses of the Tage [Teche] River," ca. 1780 (with map), Thomas Hutchins Papers, Historical Society of Pennsylvania, Philadelphia; Barthélémy Lafon, *Carte générale du territoire d'Orléans comprenant aussi la Floride Occidentale et une portion du territoire du Mississipi* (New Orleans: Barthélémy Lafon, 1806), original in the Geography and Map Division, Library of Congress, Washington, DC. Lafon's map can be viewed on the Library of Congress's American Memories website at www.memory.loc.gov/ammem/index.html.

In his petition of October 10, 1763, Masse noted that he had established a cattle ranch in Louisiana "more than sixteen years before"—i.e., 1746 or earlier. Masse's petition appears in translation in Michael James Foret, "Aubry, Foucault, and the Attakapas Acadians: 1765," *Attakapas Gazette* XV (Summer 1980): 63.

Robert S. Weddle claimed that Masse appeared in south Louisiana's historical record as early as 1728. Weddle assumed, however, that a certain "de Massy"—mentioned in genealogist Winston De Ville's *Opelousas: The History of a French and Spanish Military Post in America, 1716–1803*—was the same person as Masse. This seems unlikely: even De Ville did not consider de Massy to be the same person as Masse, noting "[I]t cannot be determined who de Massy is" (though, mused De Ville, he might have been a Louisiana resident named Jean Massy, originally of Tours, France). Robert S. Weddle, *The French Thorn: Rival Explorers in the Spanish Sea, 1682–1762* (College Station: Texas A&M University Press, 1991), 291; Winston De Ville, *Opelousas: The History of a French and Spanish Military Post in America, 1716–1803* (Cottonport, LA: Polyanthos, 1973), 26, 144 (fn. 10).

Some colonial and modern sources have unconvincingly placed Masse in southeast Texas or extreme southwest Louisiana. For the purposes of this book, however, it matters only that Masse resided at some time along the Teche, which he evidently did from the 1740s to the end of his life between 1771 and 1772.

It is unclear if Hutchins visited the Teche area himself or merely collected data from others who visited the region. Although he traveled to Louisiana in or around 1772, Hutchins's published text, the notes on which he based his text, and his hand-drawn map of the Tage (Teche) all mention "*Nouvelle Iberie*" (New Iberia), a settlement founded in 1779—thus indicating at least some of his research dated from that year or shortly afterward.

For his measurements of the Teche and adjacent landmarks, Hutchins relied on the traditional Spanish league (one league equals 2.63 miles).

12. Bolton, *Texas in the Middle Eighteenth Century*, 360.

13. Ibid., 298–99.

14. José Antonio Pichardo, *Pichardo's Treatise on the Limits of Louisiana and Texas, Vol. II*, Charles Wilson Hackett, trans., ed., and anno. (Austin: University of Texas Press, 1931),

201, 202; Pichardo, *Pichardo's Treatise on the Limits of Louisiana and Texas, Vol. I*, Charles Wilson Hackett, trans., ed., and anno. (Austin: University of Texas Press, 1931), 376, 410; Bolton, *Texas in the Middle Eighteenth Century*, 360–61; Barrios to Amarillas.

While it seems fanciful that Masse exerted "absolute dominion" over any Native American tribes, he once offered the Spanish the cooperation of the Attakapas, Koroa, Taovaya, Pawnee, and Comanche. This offer hinged on Spanish permission for Masse to resettle with his slaves at the presidio of San Agustin de Ahumada (in present-day east Texas). The Spanish refused, however, abhorring the idea of a Frenchman residing in undeniably Spanish territory. Pichardo, *Pichardo's Treatise, Vol. I*, 410.

15. Masse died prior to December 15, 1784. Proceedings regarding the succession of André Masse, in Laura L. Porteous, "Index to the Spanish Judicial Records of Louisiana, LXXIV [January 1785]," *Louisiana Historical Quarterly* 25 (July 1942): 874–76.

Donald J. Arceneaux has determined that Masse died after February 1772 but before January 1773. Donald J. Arceneaux, Moscow, Idaho, to e-mail to the author, March 17, 2015.

16. C. C. Robin, *Voyage to Louisiana*, trans. Stuart O. Landry Jr. (Gretna, LA: Firebird Press/Pelican Publishing Company, 2000), 189.

17. Ibid., 189–90; Bolton, *Texas in the Middle Eighteenth Century*, 359–60; Carl A. Brasseaux, Keith P. Fontenot, and Claude F. Oubre, *Creoles of Color in the Bayou Country* (Jackson: University Press of Mississippi, 1994), 7.

18. Vincent H. Cassidy and Mathé Allain, "The Attakapas Territory: 1721–1747," *Attakapas Gazette* III (June 1968): 15; Mathé Allain and Vincent H. Cassidy, "Blanpain, Trader among the Attakapas," *Attakapas Gazette* III (December 1968): 32–33, 36, 38; Claude [F.] Oubre, "Port Barre: A Crossroads in the Opelousas Country," *Attakapas Gazette* XI (Spring 1976): 43; Carl A. Brasseaux and Philip Gould, *Acadiana: Louisiana's Historic Cajun Country* (Baton Rouge: Louisiana State University, 2011), 63.

Oubre stated that it was Jacques-Guillaume Courtableau who established a trading post at the divergence of bayous Courtableau and Teche; but I defer to Carl A. Brasseaux, who more recently asserted that it was Le Kintrek who established the post.

19. Oubre, "Port Barre," 43; Carl A. Brasseaux, *The Founding of New Acadia: The Beginnings of Acadian Life in Louisiana, 1765–1803* (Baton Rouge: Louisiana State University Press, 1987), 98; Gertrude C. Taylor, "Land Grants along the Teche, Part I: Port Barre to St. Martinville," map [Lafayette: Attakapas Historical Association/Center for Louisiana Studies, University of Southwestern Louisiana], 1979).

20. Glenn R. Conrad, "A Lady Called Alice," *Attakapas Gazette* XIII (Fall 1978): 125–26; Emma Fuselier Philastre, "Gabriel Fuselier de la Claire," *Attakapas Gazette* IX (June 1974): 77; Taylor, "Land Grants along the Teche, Part I," map; Winston De Ville, "Fuselier de la Claire and the Lands of Attakapas and Opelousas," *Mississippi Valley Mélange, Volume One: A Collection of Notes and Documents for the Genealogy and History of the Province of Louisiana and the Territory of Orleans* (Ville Platte, LA: Winston De Ville, 1995), 36–37.

21. Carl A. Brasseaux, *French, Cajun, Creole, Houma: A Primer on Francophone Louisiana* (Baton Rouge: Louisiana State University Press, 2005), 97–98; John Mack Faragher, *A Great and Noble Scheme: The Tragic Story of the Expulsion of the French Acadians from Their*

American Homeland (New York: W. W. Norton, 2005), 430; Winston De Ville, *Mississippi Valley Mélange, Volume Two: A Collection of Notes and Documents for the Genealogy and History of the Province of Louisiana and the Territory of Orleans* (Ville Platte, LA: Winston De Ville, 1996), 14; Brasseaux and Gould, *Acadiana*, 64; Taylor, "Land Grants along the Teche, Part I," map. For more on Degoutin, see Tim Hebert, "The First Acadian in Louisiana: Joseph de Goutin de Ville," ca. 2000, Acadian-Cajun Genealogy & History, www.acadian-cajun.com/degoutin.htm.

22. William D. Reeves, *From Tally-Ho to Forest Home: The History of Two Louisiana Plantations* (Bloomington, IN: AuthorHouse, 2005), 20; Carl A. Brasseaux, *"Scattered to the Wind": Dispersal and Wanderings of the Acadians, 1755–1809*, Louisiana Life Series, No. 6 (Lafayette: Center for Louisiana Studies/University of Louisiana at Lafayette, 1991), 62–64; Brasseaux, *Founding of New Acadia*, 74, 198; Carl A. Brasseaux, *Acadian to Cajun: Transformation of a People, 1803–1877* (Jackson: University Press of Mississippi, 1992), xi, xiv; Warren A. Perrin, *Acadian Redemption: From Beausoleil Broussard to the Queen's Royal Proclamation* (Erath, LA: Warren A. Perrin, 2004), 34, 39.

Brasseaux referred to "Edouard Masse." Similarly, Edward T. Weeks Sr. referred to "Edward Masse." From context, however, they clearly meant André Masse. Brasseaux, *Founding of New Acadia*, 75; Edward T. Weeks Sr., "Some Facts and Traditions about New Iberia," in Glenn R. Conrad, ed., *New Iberia: Essays on the Town and Its People*, 2nd ed. (Lafayette: Center for Louisiana Studies, University of Southwestern Louisiana, 1986), 17.

23. Casualty statistics for the Acadian expulsion can be found in Faragher, *A Great and Noble Scheme*, 424–25, 470–71, 473.

24. Brasseaux, *Founding of New Acadia*, 74–75, 123. Brasseaux describes the cattle raised by Acadian exiles as "semidomesticated longhorns."

25. Ibid., 75.

26. Ibid., 76, 91–92, 94; Perrin, *Acadian Redemption*, 42–43; Donald J. Arceneaux, "A New Look at the Initial Acadian Settlement Location in the Attakapas," *Attakapas Gazette* (online version), revised June 10, 2015, www.attakapasgazette.org/vol-3-2014/initial-acadian-settlement/, accessed June 13, 2015, n.p.

I am indebted to Donald Arceneaux for sharing his insights about the locations of the Acadians' camps along the Teche.

Brasseaux suggested that the settlement known as La Manque sat on the Teche near Breaux Bridge, but Arceneaux has developed a convincing argument that La Manque sat on the bayou between present-day New Iberia and Loreauville.

Arceneaux and I concur that, contrary to popular belief, Joseph Broussard dit Beausoleil never resided at Côte Gelée or founded the settlement that became Broussard, Louisiana; rather, we believe Broussard dit Beausoleil resided at Camp Beausoleil on Bayou Teche up to the time of his death in October 1765.

27. Brasseaux, *Founding of New Acadia*, 92; Arceneaux, "The Initial Acadian Settlement," n.p.

28. Brasseaux, *Founding of New Acadia*, 94; Arceneaux, "The Initial Acadian Settlement," n.p.

The epidemic is also mentioned in Jean-François Mouhot, ed., "Letter by Jean-Baptiste Semer, an Acadian in New Orleans, to His Father in Le Havre, April 20, 1766," trans. Bey Grieve, *Louisiana History* 48 (Spring 2007): 224.

Spanish land grant data comes from the author's analysis of Taylor, "Land Grants along the Teche, Part I," map; Gertrude C. Taylor, "Land Grants along the Teche, Part II: St. Martinville to Sorrel," map (Lafayette: Attakapas Historical Association/Center for Louisiana Studies, University of Southwestern Louisiana, 1980); Gertrude C. Taylor, "Land Grants along the Teche, Part III: Sorrel to Berwick Bay," map (Lafayette: Attakapas Historical Association/Center for Louisiana Studies, University of Southwestern Louisiana, 1980).

29. Fontenot, "Livestock of Old Southwest Louisiana," 80; Brasseaux, *Founding of New Acadia*, 123–24; Lauren C. Post, *Cajun Sketches, from the Prairies of Southwest Louisiana* (Baton Rouge: Louisiana State University Press, 1990), 43–44; Lauren C. Post, "The Old Cattle Industry of Southwest Louisiana," *McNeese Review* (1957), reprinted in Glenda Schoeffler, ed., *Cattle Brands of the Acadians and Early Settlers of Louisiana/Attakapas* (Lafayette, LA: Self-published, 1992), 46.

30. Jack D. L. Holmes, "Indigo in Colonial Louisiana and the Floridas," *Louisiana History* VIII (Fall 1967): 338–39, 344–46.

31. Ibid., 330–31; Steve Canac-Marquis and Pierre Rézeau, "Introduction," *Journal de Vaugine de Nuisement: un témoignage sur la Louisiane du XVIIIe siècle* (Quebec City: Éditions Presse de l'Université de Laval, 2005), 3. See also Gertrude Taylor, "Etienne de Vaugine: Soldier, Planter, Trader," *Attakapas Gazette* XV (Summer 1980): 53–59.

32. Canac-Marquis and Pierre Rézeau, "Introduction," 3, 5; Taylor, "Land Grants along the Teche, Part II," map; Henry P. Dart, "A Louisiana Indigo Plantation on Bayou Teche, 1773," *Louisiana Historical Quarterly* 9 (October 1926): 566–67; Laura L. Porteous, trans., "Inventory of de Vaugine's Plantation in the Attakapas on Bayou Teche, 1773," *Louisiana Historical Quarterly* 9 (October 1926): 570–71, 580. [Note: I have drawn on both Porteous's translation of the de Vaugine inventory and the untranslated transcript following it.]

33. Canac-Marquis and Rézeau, "Introduction," 3–4, 5; Dart, "Louisiana Indigo Plantation," 566; Porteous, "Inventory of de Vaugine's Plantation," 570, 580.

34. Porteous, "Inventory of de Vaugine's Plantation," 574–77, 583–86.

35. Ibid., 571–72, 577, 580–81, 586–87; Shane K. Bernard, *Swamp Pop: Cajun and Creole Rhythm and Blues* (Jackson: University Press of Mississippi, 1996), 88; Ann Allen Savoy, comp. and ed., *Cajun Music: A Reflection of a People, Vol. 1*, 3rd ed. (Eunice, LA: Bluebird Press, 1988), 108.

Porteous translated "*traîneaux*" as "drag-nets," but in Louisiana French the word denotes drag sleds; *Dictionary of Louisiana French*, s.v. "*traîneau*."

36. Porteous, "Inventory of de Vaugine's Plantation," 578–79, 587–588.

37. Ibid., 572–73, 578, 582–83, 587. In her *Attakapas Gazette* article, Taylor counted thirty-two de Vaugine slaves, while in his *Louisiana Historical Quarterly* article Dart counted thirty-three. I am unsure how Taylor arrived at her total, but Dart apparently excluded the two runaways. I include the two runaways in my tally, however, because de Vaugine's inventory counted them.

38. Porteous, "Inventory of de Vaugine's Plantation," 571, 573, 581, 583.

The "*barre a prisonnier*" referred to in the de Vaugine inventory is undoubtedly the same as the "*barres de prisonniers*" mentioned under the entry for "Bilboes" in William Falconer and William Burney, *Falconer's New Universal Dictionary of the Marine* (1815; reprint, Annapolis, MD: Naval Institute Press, 2006). Always used in plural, *bilboes* is the English word for "*barres de prisonniers*." *Falconer's* defines *bilboes* as "long bars or bolts of iron, with shackles sliding on them, and a lock at the end, used, on some occasions, to confine the feet of prisoners, in a manner similar to the confinement of the hands in handcuffs."

39. Jean-Francois Ledée to Alexandre DeClouet, December 21, 1778, Legajo 198A, 176, Archive General de Indias, Papeles Procedentes de Cuba, in research notes of Carl A. Brasseaux, Lafayette, LA; "Judge Jehu Wilkinson's Reminiscences," *Attakapas Gazette* XI (Fall 1976): 141; Entry for Jehu Wilkinson, Find A Grave, www.findagrave.com, accessed February 14, 2015.

40. Barry Ancelet, Jay Edwards, and Glen Pitre, *Cajun Country* (Jackson: University Press of Mississippi, 1991), 61; Richardson, "The Teche Country Fifty Years Ago," 119; Holmes, "Indigo in Colonial Louisiana," 340, 347–49.

41. Gilbert C. Din, "Lieutenant Colonel Francisco Bouligny and the Malagueño Settlement at New Iberia, 1779," *Louisiana History* XVII (Spring 1976): 187–88, 190, 191, 192–93; Gilbert C. Din, *Francisco Bouligny: A Bourbon Soldier in Spanish Louisiana* (Baton Rouge: Louisiana State University Press, 1993), 2, 3–4, 12, 29–30, 31–39, 40–41, 43, 67–71, 75–77, 83, 96. Din estimated the number of Malagueños who arrived in Louisiana at "over sixty."

42. Din, "Bouligny and the Malagueño Settlement," 192; Din, *Bourbon Soldier in Spanish Louisiana*, 92–96.

43. Din, "Bouligny and the Malagueño Settlement," 193; Din, *Bourbon Soldier in Spanish Louisiana*, 96–97; Allain, "Bouligny's Account [part one]," 81–82. I rely primarily on Allain's translations of Bouligny's reports about the founding of Nueva Iberia; Bergerie, however, translated some of the same documents. See Bergerie, *They Tasted Bayou Water*, 131–40.

44. Francisco Bouligny, Nueva Iberia, to Bernardo de Gálvez, New Orleans, February 18, 1779, ADS, Legajo 2358, microfilm copy in Louisiana Room, Edith Garland Dupré Library, University of Louisiana at Lafayette; Din, *Bourbon Soldier in Spanish Louisiana*, 97; Allain, "Bouligny's Account [part one]," 83. The quote is my own translation from the original document.

45. Allain, "Bouligny's Account [part one]," 83; Melanie Marcotte, "News from the Past: Chitimacha Chiefs," *Chitimacha Newsletter* [publication of the Sovereign Nation of the Chitimacha], August 1997, 5.

46. Allain, "Bouligny's Account [part one]," 81, 83.

47. Din, "Bouligny and the Malagueño Settlement," 195; Mathé Allain, "Bouligny's Account of the Founding of New Iberia [part two]," *Attakapas Gazette* XIV (Fall 1979): 125–26.

48. Din, "Bouligny and the Malagueño Settlement," 195; Din, *Bourbon Soldier in Spanish Louisiana*, 98; Allain, "Bouligny's Account [part two]," 124, 125.

49. Din, "Bouligny and the Malagueño Settlement," 195–96; Din, *Bourbon Soldier in Spanish Louisiana*, 98; Allain, "Bouligny's Account [part two]," 127.

50. Allain, "Bouligny's Account [part two]," 127.

51. Ibid.; Bergerie, *They Tasted Bayou Water*, 7 (fn. 3); Francisco Bouligny, Nueva Iberia, to Bernardo de Gálvez, New Orleans, 30 June 1779, ADS, Legajo 2358, microfilm copy in Louisiana Room, Edith Garland Dupré Library, University of Louisiana at Lafayette.

52. Allain, "Bouligny's Account [part two]," 127; Din, "Bouligny and the Malagueño Settlement," 196.

Chapter 2

1. Andrew Sluyter, "The Role of Blacks in Establishing Cattle Ranching in Louisiana in the Eighteenth Century," *Agricultural History* 86 (Spring 2012): 43, 45.

2. David K. Bjork, "Documents Relating to Alexandro O'Reilly and an Expedition Sent out by Him from New Orleans to Natchitoches, 1769–1770," *Louisiana Historical Quarterly* VII (January 1924): 23–26.

Kelly is identified as an artillery lieutenant in Bjork, "Documents Relating to Alexandro O'Reilly," 23 (fn. 9). Nugent is referred to as an infantry officer in *Gazetta de Madrid*, November 14, 1775, 407.

3. Bjork, "Documents Relating to Alexandro O'Reilly," 27–30, 35, 36, 38.

4. Joseph G. Tregle Jr., "British Spy along the Mississippi: Thomas Hutchins and the Defenses of New Orleans, 1773," *Louisiana History* VIII (Fall 1967): 314–15, 326–27.

5. Ibid., 315.

6. Joseph G. Tregle Jr., "Introduction," in Thomas Hutchins, *An Historical Narrative and Topographical Description of Louisiana, and West-Florida* [*sic*] (Philadelphia: Robert Aitken, 1784), v–vi, xliii.

7. Berwick and Henderson helped Bouligny establish New Iberia. See Michael James Foret, "The Berwicks of St. Mary Parish," *Attakapas Gazette* XXI (Spring 1986): 3; Glenn R. Conrad, "Some Observations on Anglo-Saxon Settlers in Colonial Attakapas," *Attakapas Gazette* XX (Spring 1985): 42. Conrad refers to over a dozen "Anglo-Saxons" residing in New Iberia or its environs by 1781. See Conrad's annotation to Dr. Alfred Duperier, "The Obituary of William F. Weeks," in Glenn R. Conrad, ed., *New Iberia: Essays on the Town and Its People*, 2nd ed. (Lafayette: Center for Louisiana Studies, University of Southwestern Louisiana, 1986), 115–18 (fn. 4).

8. Hutchins, *An Historical Narrative*, 47; Hutchins, map of the Apelousa River, Thomas Hutchins Papers, Historical Society of Pennsylvania, Philadelphia. Lorimer is mentioned in Peter Wilson Coldham, ed., *North American Wills Registered in London, 1611–1857* (Baltimore, MD: Genealogical Publishing, 2007), 38. Among other connections, Hutchins and Lorimer both belonged to the American Philosophical Society. See Christopher P. Iannini, *Fatal Revolutions: Natural History, West Indian Slavery, and the Routes of American Literature* (Chapel Hill: University of North Carolina Press, 2012), 183. I credit archaeologist Donny Bourgeois with identifying Dr. John Lorimer.

9. Hutchins, 46–48. Hutchins's hand-drawn map of the Teche can be found in the Thomas Hutchins Papers, Historical Society of Pennsylvania, Philadelphia.

10. Jack D. L. Holmes, "Evia, Jose Antonio de," *Handbook of Texas Online* [Texas State Historical Association], www.tshaonline.org/handbook/online/articles/fev13, accessed November 21, 2011.

11. José de Evia, *José de Evia y sus reconocimientos del Golfo de México, 1783–1796*, ed. Jack D. L. Holmes (Madrid: Ediciones José Porrúa Turanzas, 1968), 87, 91–92, 115–16, 118; Taylor, "Land Grants along the Teche, Part I," map.

12. Evia, *José de Evia*, 126, 134–35.

13. Ibid., 92–93, 116, 118. *Berchas* are described in W. Raymond Wood, *Prologue to Lewis and Clark: The Mackay and Evans Expedition* (Norman: University of Oklahoma Press, 2003), 67–68.

14. Evia, *José de Evia*, 93–94, 116, 118; Robert S. Weddle, *Changing Tides: Twilight and Dawn in the Spanish Sea, 1763–1803* (College Station: Texas A&M University Press, 1995), 153–57, 186.

15. Evia, *José de Evia*, 99–101, 102, 107–8, 120, 129–30, 135, 141.

16. Unless otherwise stated, all information about Gonsoulin and Grevemberg's exploration of the Teche derives from François Gonsoulin, "Routties fait de la Nouvelle Iberie à la sortie du Thex, et retour à la ditte par mer par la petite ance [*sic*]," June 18–28, 1779, ADS, Legajo 2358, Papeles Procedentes de Cuba, Archivo General de Indias, Seville, Spain, microfilm copy in the Louisiana Room, Edith Garland Dupré Library, University of Louisiana at Lafayette. Francisco Bouligny, Nueva Iberia, to Bernardo de Gálvez [New Orleans?], June 30, 1779, ADS, Legajo 2358, Papeles Procedentes de Cuba, Archivo General de Indias, Seville, Spain, microfilm copy in the Louisiana Room, Edith Garland Dupré Library, University of Louisiana at Lafayette.

17. Jean-Baptiste Grevemberg, "Journal des relevé des criques, rivières, bayes, isles, etc., depuis la Rivière Teiche [*sic*]," July 18–25, 1779, ASD [microfilm], Legajo 2358, Papeles Procedentes de Cuba, Archivo General de Indias, Seville, Spain, microfilm copy in the Louisiana Room, Edith Garland Dupré Library, University of Louisiana at Lafayette.

18. Rev. Donald J. Hébert, *Southwest Louisiana Records, Vol. 1-B (1801–1810)* (Rayne, LA: Hébert Publications, 1996), 324–25; Bouligny to Gálvez, June 30, 1779. Gonsoulin provided a brief autobiography at the end of "Routties fait de la Nouvelle Iberie."

19. "Cuenta de los gastos ocasionados para los tres viages que ha hecho Dn. Juan Bauta. Grevember [*sic*]," August 26, 1779, ADS, Documentos de la Luisiana (MSS 19248), folio 141, Sección de Manuscritos, Biblioteca Nacional, Madrid, Spain; Bouligny to Gálvez, June 30, 1779. See also Evia, *José de Evia y sus reconocimientos*, 91–92 (fn. 114), which refers to this same document and the pages adjacent to it as "el reconocimiento por Juan Bautista Grevembert [*sic*]."

20. Bouligny refers to the Lower Atchafalaya as "*el Rio Grande*" in Francisco Bouligny, Nueva Iberia, to Bernardo de Gálvez [New Orleans?], May 16, 1779, and June 30, 1779, ADS, Legajo 2358, Papeles Procedentes de Cuba, Archivo General de Indias, Seville, Spain, microfilm copy in Louisiana Room, Edith Garland Dupré Library, University of Louisiana at Lafayette.

21. The modern *Louisiana Atlas & Gazetteer* published by DeLorme still associates a "red bluff" with Côte Blanche Island; *Louisiana Atlas & Gazetteer* (Yarmouth, ME: DeLorme, 2003), 54.

22. Shane K. Bernard, "Gumbo in 1764?" *Bayou Teche Dispatches*, blog, October 3, 2011, www.bayoutechedispatches.blogspot.com/2011/10/gumbo-in-1764.html, accessed February 28, 2015. Historian Gwendolyn Midlo Hall found this early gumbo reference in the records of the French Superior Council, Louisiana Historical Center, New Orleans.

23. Bouligny to Gálvez, June 30, 1779.

24. Shane K. Bernard, "Avery Island," *KnowLA Encyclopedia of Louisiana*, www.knowla.org/entry/1210/&view=article, accessed April 10, 2013.

25. *Gonsoulin's Heirs v. Gulf Co.*, No. 1,131, Circuit Court of Appeals, Fifth Circuit, May 27, 1902, in *Federal Reporter* 116 (1902): 252. This source states that Gonsoulin received his Belle Isle grant from the Spanish in 1783.

Chapter 3

1. Din, *Bourbon Soldier in Spanish Louisiana*, 101; Winston De Ville, *Louisiana Soldiers in the American Revolution* (Ville Platte, LA: Provincial Press, 1991), 23–26; Roster of names, monument to American Revolution soldiers buried at St. Martin de Tours Catholic Church, St. Martinville, LA; John Spencer Bassett, ed., *Correspondence of Andrew Jackson: to April 30, 1814, Vol. 2* (Washington: Carnegie Institution, 1937), 169; Marion John Bennett Pierson, comp., *Louisiana Soldiers in the War of 1812* (Baton Rouge: Louisiana Genealogical and Historical Society, 1963; repr. Baltimore: Clearfield, 2003), passim.

De Ville lists the members of DeClouet's unit in 1777, two years before their wartime duty. I thus defer to the list of DeClouet's soldiers engraved on the monument in St. Martinville, which is similar but not identical to De Ville's list.

2. Glenn R. Conrad, "Henry Hopkins Raises the U.S. Flag over the Attakapas," *Attakapas Gazette* XXII (Summer 1987): 50.

3. The late-eighteenth-century presence of German and Irish settlers on the Teche is noted in Conrad, "Anglo-Saxon Settlers in Colonial Attakapas," 42; and Glenn R. Conrad, "Friend or Foe? Religious Exiles at the Opelousas Post in the American Revolution," *Attakapas Gazette* XII (Fall 1977): 137; Dennis Gibson, ed. and anno., "The Journal of John Landreth," Pt. I, *Attakapas Gazette* XIV (Fall 1979): 105.

4. Conrad, "Anglo-Saxon Settlers in Colonial Attakapas," 43, 44.

5. Ibid., 44.

6. Ibid., 44–46; Conrad, "Friend or Foe?" 137–40. See also Glenn R. Conrad and Gertrude C. Taylor, "Virginians in the Teche Country," *Attakapas Gazette* XVII–XVIII (Spring 1982–Spring 1983).

7. Dolores Egger Labbé, "Anglos in Antebellum Attakapas and Opelousas," *Attakapas Gazette* XXI (Spring 1986): 15.

8. D. D. T. Leech, *Post Office Directory; or Business Man's Guide to the Post Offices in the United States* (New York: J. H. Colton, 1856), 73.

9. Glenn R. Conrad, "*In the Beginning . . .* The Origins of St. Martinville," *Attakapas Gazette* (Yearbook 1994): 1, 4–7, 16, 22; *Territorial Papers*, 677; *Acts Passed at the First Session*

of the Third Legislature of the State of Louisiana (New Orleans: J. C. de St. Romes, 1817), 50–56; An Emigrant from Maryland [*sic*], "Views of Louisiana," *Niles' Weekly Register*, September 13, 1817, 40.

The 1769 census of the Attakapas District shows six or seven people living on Dauterive's vacherie (the future site of St. Martinville), including two *engagés* and four or five other persons thought to be slaves. Donald J. Arceneaux, *Attakapas Post in 1769: The First Nominal Census of Colonial Settlers in Southwest Louisiana* (Baton Rouge: Provincial Press/Claitor's, 2014), 25. The 1771 census of the Attakapas District shows four persons living on Dauterive's vacherie, including three black (presumably enslaved) residents and one non-black resident. Winston De Ville, trans. and ed., *Attakapas Post: The Census of 1771* (Ville Platte, LA: Provincial Press, 1986), 11.

10. Glenn R. Conrad, "New Iberia: The Spanish Years," in Glenn R. Conrad, ed., *New Iberia: Essays on the Town and Its People,* 2nd ed. (Lafayette: Center for Louisiana Studies, University of Southwestern Louisiana, 1986), 8, 10, 12–13.

11. Glenn R. Conrad, "John Stine: Early Sheriff of Attakapas," *Attakapas Gazette* XXII (Fall 1987): 112; Duperier, "Obituary of William F. Weeks," 115 (fn. 2); Dennis Gibson, ed. and anno., "The Journal of John Landreth," Pt. II, *Attakapas Gazette* XV (Summer 1980): 78.

In 1851 Dr. Alfred Duperier recorded that the population of New Iberia "does not exceed 250 inhabitants." A. Duperier, M.D., "Description of an Epidemic Bilious Remittent," *New Orleans Medical and Surgical Journal* VII (March 1851): 574.

12. Gertrude C. Taylor, "A Town Named Franklin," *Attakapas Gazette* XXI (Fall 1986): 99–103, 105, 107; Gibson, "Journal of John Landreth," Pt. I, 104–5.

13. Glenn R. Conrad and Ray F. Lucas, *White Gold: A Brief History of the Louisiana Sugar Industry, 1795–1995* (Lafayette: Center for Louisiana Studies, University of Southwestern Louisiana, 1995), 4; Post, *Cajun Sketches*, 71.

14. Bjork, "Documents Relating to Alexandro O'Reilly," 39; Brasseaux, *Founding of New Acadia*, 136–37; Carl A. Brasseaux, ed., "An 1810 Census Report on the State of Manufacturing in the Northeastern Section of the Attakapas District," *Attakapas Gazette* X (Summer 1975): 102.

15. Brasseaux, "An 1810 Census Report," 102; "Judge Jehu Wilkinson's Reminiscences," 141.

16. "Judge Jehu Wilkinson's Reminiscences," 141.

17. Ibid., 141; Mouhot, "Letter by Jean-Baptiste Semer," 225; H. C. Collins, "Report of H. C. Collins, Assistant Engineer," *Annual Report of the Chief of Engineers, United States Army, to the Secretary of War, for the Year 1883, Pt. II* (Washington: U.S. Government Printing Office, 1883), 1114; Glenn R. Conrad, anno., "Sale of an Early Sugar Plantation in the Teche Country," *Attakapas Gazette* XXVIII (Yearbook 1993): 22–26.

The superintendent of Keystone Plantation, John Peters, stated in 1899 that "Many years ago [1845 or earlier] the water in the [Keystone] canal was carried by a flume down near the sugar house and was utilized to run a water wheel near the Teche for a sawmill and to grind corn and to even grind cane at the sugar house." See "Mr. John Peters

Disclaims Lowering the Level of Spanish Lake," *New Iberia Enterprise*, November 18, 1899, 2.

The Louisiana Planter and Sugar Manufacturer identified the builder of the water-powered mill at Keystone Plantation as "C. D. and Despanie DeBlanc." "C. D.," however, is evidently a mangling of "St. Denis," and "Despanie" of "Despanet," for "St. Denis Despanet DeBlanc" was the actual name of a former owner of the plantation. In another issue of the journal the same author (or someone using the same pseudonym) correctly identified the builder of that water-powered mill as "St. Denis DeBlanc." See Perambulator [pseudonym], "Attakapas Letter," *Louisiana Planter and Sugar Manufacturer*, October 27, 1888, 197; Perambulator, "Attakapas," *Louisiana Planter and Sugar Manufacturer*, April 9, 1892, 257. See also *Annual Report of the Board of Swamp Land Commissioners to the Legislature of the State of Louisiana, January 1860* (Baton Rouge: J. M. Taylor, 1860), 111.

Some sources credit Louis-Charles De Blanc with building the watermill on the Teche; Louis-Charles was the grandfather of St. Denis DeBlanc.

For information about de Boré's impact on the Louisiana sugar industry, see Conrad and Lucas, *White Gold*, 5–7; René J. Le Gardeur Jr., "The Origins of the Sugar Industry in Louisiana," *Green Fields: Two Hundred Years of Louisiana Sugar* (Lafayette: Center for Louisiana Studies, University of Southwestern Louisiana, 1980), 10–22.

18. François-Xavier Martin, *The History of Louisiana, from the Earliest Period, Vol. 1* (New Orleans: Lyman and Beardslee, 1827), lxxxii; Richardson, "The Teche Country Fifty Years Ago," 126.

19. Data about the number of sugar planters along the Teche in 1844 comes from my analysis of P. A. Champomier, "Statement of Sugar Made in Louisiana in 1844," *Senate Documents, 62d. Cong., 1st Sess., 4 April–22 August 1911, Vol. 10: Tariff Proceedings and Documents, 1839–1857, Pt. 3* (Washington: U.S. Government Printing Office, 1911), 1929–32. I exclude from my analysis three sugar planters of unclear ethnicity.

I have tried to confirm the race and ethnicity of these sugar planters by cross-referencing Champomier's list of planters with those identified in the 1850 U.S. Census for St. Martin and St. Mary parishes, Louisiana. Unlike previous decennial censuses, the 1850 Census listed the name of every person in a household, as well as their race and place of birth.

20. Information on the Senette and Verdun families derives from the 1850 U.S. Census, St. Mary Parish, Louisiana, and 1850 U.S. Census Slave Schedule, St. Mary Parish, Louisiana. See also Brasseaux, *Creoles of Color in the Bayou Country*, 75.

21. Conrad and Lucas, *White Gold*, 19–21, 25–28; *Dictionary of Louisiana French*, s.v. "*roulaison*."

22. Richardson, "The Teche Country Fifty Years Ago," 123.

23. Conrad and Lucas, *White Gold*, 21–22; Champomier, "Statement of Sugar Made in Louisiana in 1844," 1929–32; Brasseaux and Fontenot, *Steamboats on Louisiana's Bayous*, 18–19, 19–20, 22, 39 (fn. 6); "Report of Mr. J. A. Hayward, Assistant Engineer," January 23, 1875, *Annual Report of the Chief of Engineers to the Secretary of War for the Year 1875, Pt. I* (Washington: U.S. Government Printing Office, 1875), 884.

24. Milton B. Newton Jr., ed., *The Journal of John Landreth, Surveyor* (Baton Rouge: Geoscience Publication/Department of Geography and Anthropology, Louisiana State University, 1985), 124; Darby, *Geographical Description of the State of Louisiana*, 142; Walter Prichard, Fred B. Kniffen, and Clair A. Brown, eds., "Southern Louisiana and Southern Alabama in 1819: The Journal of James Leander Cathcart," *Louisiana Historical Quarterly* 28 (July 1945): 903, 905; "For Attakapas" (advertisement for the *James Lawrence*), (New Orleans) *Louisiana Gazette*, October 28, 1818, 2, typewritten transcript in WPA collection, State Library of Louisiana, Baton Rouge; Samuel Burch, comp., *General Index to the Laws of the United States of America from March 4th, 1789, to March 3d* [*sic*], *1827* (Washington: William A. Davis, 1828), 113; T. B. Thorpe, "Sugar and the Sugar Region of Louisiana," *Harper's New Monthly Magazine* VII (November 1853): 751.

The *James Lawrence* was capable of carrying "60 Hhds [hogsheads of sugar]"; because a hogshead weighs about 1,100 pounds, the vessel therefore could carry about 66,000 pounds, or roughly 33 short tons, of sugar.

Poling, warping, cordelling, and bushwhacking are described in Robert Gudmestad, *Steamboats and the Rise of the Cotton Kingdom* (Baton Rouge: Louisiana State University Press, 2011), 11–12; Hunt Janin, *Claiming the American Wilderness: International Rivalry in the Trans-Mississippi West, 1528–1803* (Jefferson, NC: McFarland, 2006), 167; J. Thomas Scharf, *History of Saint Louis City and County, Vol. II* (Philadelphia: Louis H. Everts, 1883), 1090; Lylie O. Harris, "Creole Cookery, No. IV: A Glimpse at Home Life on the Têche," *Home-Maker* III (January 1890): 327.

25. Brasseaux and Fontenot, *Steamboats on Louisiana's Bayous*, 68–69; Gibson, "Journal of John Landreth," Pt. II, 78; "Extract of a [L]etter from the Collector of Nova Iberia to the Secretary of the Treasury, Dated at Baltimore, December 10, 1817," *Accompanying the Bill to Authorize the Building of a Certain Number of Small Vessels* (Washington: Gales & Seaton, 1820), 8–9; "Letters from the South" [New Orleans, February 24 (1855)], *Western Literary Messenger* XXIV (April 1855): 89; Duperier, "Obituary of William F. Weeks," 128–29, see also fn. 25 on same pages.

For a sample of Teche Country tales about Lafitte, see Patrick Flanagan, "A Pirate's Tale," (New Iberia, LA) *Daily Iberian*, May 16, 2011, www.iberianet.com/people/features/a-pirate-s-tale/article_2e597d5e-7fe5-11e0-9735-001cc4c002e0.html, accessed June 14, 2015.

26. Richard Peters, ed., *The Public Statutes at Large of the United States of America, Vol. III* (Boston: Charles C. Little and James Brown, 1848), 392; "Appointment by the President, by and with the Advice and Consent of the Senate," *Milwaukee Sentinel and Gazette*, July 24, 1846, 4; "Statement of Exports, by Sea, out of the State [of Louisiana], from the Port of Franklin, District of [the] Teche, . . . from the 30th of September, 1842, to the 30th of June, 1843" and "Statement of the Number of Vessels, Outward and Inward, at the Port of Franklin [during the same time period]," in John Macgregor, *Commercial Tariffs and Regulations, Resources, and Trade, Pt. 15, United States of America* (London: Charles Whiting, 1846), 462.

It is unclear if Macgregor's "Statement of the Number of Vessels . . . at the Port of Franklin" counts only seafaring vessels, or includes inland steamboats. The chart is paired, however, with "Statement of Exports, by Sea . . . from the Port of Franklin," which clearly counts only seafaring vessels, suggesting that "Statement of the Number of Vessels . . . at the Port of Franklin" does likewise.

27. Brasseaux and Fontenot, *Steamboats on Louisiana's Bayous*, 2, 4–5, 12, 37–38.

28. Ibid., 38–39, 40 (fn. 8).

29. Ibid., 41, see also fn. 11 on same page.

30. Ibid., 14–15, 50 (Table 50), 54; "Steam Tonnage of the United States," *De Bow's Review* (May 1858): 452.

31. Author's analysis of steamships listed in "Appendix A: Some Steamboats, Keelboats, and Barges Known to Have Operated in the Bayou Country," in Brasseaux and Fontenot, *Steamboats on Louisiana's Bayous*, 153–254; Laurence Oliphant, *Patriots and Filibusters, or Incidents of Political and Exploratory Travel* (Edinburgh and London: William Blackwood and Sons, 1860), 153.

Using Brasseaux and Fontenot's list, I counted 122 steamboats that operated at least briefly on Bayou Teche between 1820 and April 1861. This figure should be considered an approximation, however, because Brasseaux and Fontenot do not purport to have compiled a complete list. Moreover, I excluded from my count steamships listed merely as "engaged with the St. Martinville trade" because such trade may have occurred not via Bayou Teche, but via Bayou Portage through the Atchafalaya swamp. Likewise, I excluded steamships listed merely as "engaged in the Attakapas trade" because this does not necessarily denote trade on Bayou Teche.

I used the distance measurement tool on Google Maps (www.maps.google.com) to gauge the width of the Teche on aerial photographs.

32. Brasseaux and Fontenot, *Steamboats on Louisiana's Bayous*, 48–52, 153–254; Shane K. Bernard, *Tabasco: An Illustrated History* (Avery Island, LA: McIlhenny Company, 2007), 37–38, 42.

Judge John Moore is quoted in *Proceedings of the New Orleans, Algiers, Attakapas and Opelousas Railroad Convention* (New Orleans: *The Orleanian*, 1851), 21; see also "The Late Opelousas Railroad Convention at New Orleans," *DeBow's Review* XI (December 1851), 670.

In the spirit of full disclosure, the author wishes to note that he is employed as historian and curator to McIlhenny Company, maker of Tabasco brand pepper sauce since 1868. He thought it important, however, to note the vital role of the Teche in introducing this cultural and culinary icon to the world.

33. See the description of snag boats in John Macgregor, *The Progress of America, from the Discovery by Columbus to the Year 1846, Vol. II: Geographical and Statistical* (London: Whittaker, 1847), 688–89. Jean Jules Hardy, Bayou Teche, St. Martin Parish, LA, to George W. Morse, State Engineer, [Baton Rouge, LA?], April 25, 1853 [handwritten copy by Hardy], Jean Jules Hardy Snag Boat Journal, original in the possession of Dr. Florent Hardy Jr., Louisiana State Archivist, Baton Rouge, LA. Dr. Hardy is the great-great-grandson of Jean Jules Hardy and the journal is presently in his personal possession.

34. Hardy Snag Boat Journal, July 25, 1853.

35. Jo Ann Carrigan, *The Saffron Scourge: A History of Yellow Fever in Louisiana, 1796–1905* (Lafayette: Center for Louisiana Studies, University of Southwestern Louisiana, 1994), 4, 7–8.

36. Ibid., 17–19, 21. Between July and November 1765, thirty-nine burial services were performed among the Acadian exiles who arrived along Bayou Teche under Joseph Broussard dit Beausoleil. It is unknown if the epidemic caused all these deaths, but the next year exile Jean-Baptiste Semer recorded, "[E]verybody contracted fevers at the same time and nobody being in a state to help anyone else, thirty-three or thirty-four died, including the children." Perrin, *Acadian Redemption*, 34, 42; Mouhot, "Letter by Jean-Baptiste Semer," 224.

37. Carrigan, *The Saffron Scourge*, 1, 5, 51, 393 (fn. 71), 396 (fn. 42).

38. Ibid., 52, 56 (table 3); Glenn R. Conrad, "New Iberia and Yellow Fever: Epidemic and Quarantine," in Glenn R. Conrad, ed., *New Iberia: Essays on the Town and Its People*, 2nd ed. (Lafayette: Center for Louisiana Studies, University of Southwestern Louisiana, 1986), 175; William Henry Perrin, ed., *Southwest Louisiana Biographical and Historical* (New Orleans: Gulf Publishing, 1891), 108–9.

39. Carrigan, *The Saffron Scourge*, 58–80, 81 (table 5); E. D. Fenner, M.D., "Report on the Epidemics of Louisiana, Mississippi, Arkansas, and Texas, in the Year 1853," *Transactions of the American Medical Association* VII (1854): 505–6.

40. Fenner, "Report on the Epidemics of Louisiana," 508. Population figure for Pattersonville is from the 1850 U.S. Census; see J. D. B. DeBow, *Statistical View of the United States* (Washington: A. O. P. Nicholson, 1854), 374.

41. Fenner, "Report on the Epidemics of Louisiana," 502–3.

42. Ibid., 504–5; Brasseaux and Fontenot, *Steamboats on Louisiana's Bayous*, 70. Population figure for Franklin is for 1853; see DeBow, *Statistical View of the United States*, 354 (fn. h). According to Dr. Fenner, the two quarantine stations below Franklin sat "near the junction of the [Lower] Atchafalaya and Teche" and "on the [Lower] Atchafalaya, a short distance below Pattersonville."

43. "The Health of Our Town," *St. Martinsville Constitutional*, reprinted in *Opelousas Courier*, September 29, 1855, 2.

Chapter 4

1. John D. Winters, *The Civil War in Louisiana* (Baton Rouge: Louisiana State University Press, 1991), 159–62.

2. Ibid., 212, 222, 298, 326.

3. U.S. War Department, *The War of the Rebellion: A Compilation of the Official Records of the Union and Confederate Armies* (hereafter *OR*), 128 vols. (Washington: U.S. Government Printing Office, 1880–1901), Series I, Vol. 15, 726; J. W. De Forest, "Forced Marches," *The Galaxy* V (June 1868): 708; Donald S. Frazier, *Fire in the Cane Field: The Federal Invasion of Louisiana and Texas, January 1861-January 1863* (Buffalo Gap, TX: State House Press, 2009), 141; Donald S. Frazier, *Thunder across the Swamp: The Fight for the Lower Mississippi, February 1863-May 1863* (Buffalo Gap, TX: State House Press, 2011), 405.

4. U.S. Naval War Records Office, *Official Records of the Union and Confederate Navies in the War of the Rebellion* (hereafter *ORN*), 27 vols. (Washington: U.S. Government Printing

Office, 1894–1922), Series I, Vol. 16, 626, Series I, Vol. 19, 327, 519–20, Series II, Vol. 1, 251; Frazier, *Fire in the Cane Field*, 210; Morris Raphael, *The Battle in the Bayou Country* (Detroit: Harlo, 1990), 57; Morris Raphael, *A Gunboat Named Diana* (Detroit: Harlo, 1993), 70.

5. *ORN*, Series I, Vol. 19, 327; Frazier, *Fire in the Cane Field*, 210–13; Robert M. Browning Jr., *Lincoln's Trident: The West Gulf Blockading Squadron during the Civil War* (Tuscaloosa: University of Alabama Press, 2015), 233; Raphael, *Battle in the Bayou Country*, 57–58; Raphael, *Gunboat Named Diana*, 70–73.

6. *ORN*, Series I, Vol. 19, 327–28; Frazier, *Fire in the Cane Field*, 213–14; Raphael, *Battle in the Bayou Country*, 58.

7. Elias P. Pellet, *History of the 114th Regiment, New York State Volunteers* (Norwich, NY: Telegraph and Chronicle Power Press, 1866), 64; James K. Hosmer, *The Color-Guard: Being a Corporal's Notes of Military Service in the Nineteenth Army Corps* (Boston: Walker, Wise, 1864), 163; Frazier, *Fire in the Cane Field*, 64, 316.

8. *ORN*, Series I, Vol. 19, 519–20, 524–25; Frazier, *Fire in the Cane Field*, 318–20, 322–23; George N. Carpenter, *History of the Eighth Regiment Vermont Volunteers, 1861–1865* (Boston: Deland & Barta, 1886), 82; Raphael, *Battle in the Bayou Country*, 69–70, 72; Raphael, *Gunboat Named Diana*, 89–90, 92–95.

9. *OR*, Series I, Vol. 24, Pt. 3, 182; Frazier, *Thunder across the Swamp*, 149; Cecil D. Eby Jr., *A Virginia Yankee in the Civil War: The Diaries of David Hunter Strother* (Chapel Hill: University of North Carolina Press, 1998), 165–66; Raphael, *Battle in the Bayou Country*, 79–83; Raphael, *Gunboat Named Diana*, 105–11.

10. Frazier, *Thunder across the Swamp*, 151, 159–60, 162, 165, 166, 168–70, 172; Raphael, *Battle in the Bayou Country*, 92. Troop strength figures for Taylor's and Banks's forces are from Christopher G. Peña, *Scarred by War: Civil War in Southeast Louisiana* (Bloomington, Ind.: AuthorHouse, 2004), 219.

11. Frazier, *Thunder across the Swamp*, 119, 173–75, 180–86, 188, 192, 216, 217, 228; Raphael, *Battle in the Bayou Country*, 95–99; Raphael, *Gunboat Named Diana*, 117–36; Winters, *Civil War in Louisiana*, 222–23, 225; Richard Taylor, *Destruction and Reconstruction: Personal Experiences of the Late War* (New York: D. Appleton and Company, 1879), 131; De Forest, "Forced Marches," 710.

12. Frazier, *Thunder across the Swamp*, 183, 245–46, 249 (map), 250–54, 256–57, 262, 276–77, 279–81, 286, 292, 294–98, 303, 308–10, 312; Raphael, *Battle in the Bayou Country*, 106–7, 111–13, 116–17, 123; Winters, *Civil War in Louisiana*, 226; Taylor, *Destruction and Reconstruction*, 133; Richard Biddle Irwin, *History of the Nineteenth Army Corps* (New York: G. P. Putnam's Sons, 1892), 107; Calvin Stebbins, *Henry Hill Goodell: The Story of His Life with Letters and a Few of His Addresses* (Cambridge: Riverside Press, 1911), 40–41; H. M. Whitney, "Up the Teche with General Banks," in W. C. King and W. P. Derby, comps., *Camp-Fire Sketches and Battle-Field Echoes* (Springfield, MA: King, Richardson, 1886), 163; Hosmer, *The Color-Guard*, 135; John Farwell Moors, *History of the Fifty-Second Regiment, Massachusetts Volunteers* (Boston: George H. Ellis, 1893), 117.

13. Frazier, *Thunder across the Swamp*, 98, 321, 325–26, 327, 340–41; Raphael, *Battle in the Bayou Country*, 117, 132; Raphael, *Gunboat Named Diana*, 159; Harris H. Beecher, *Record*

of the 114th Regiment, N.Y.S.V.: Where It Went, What It Saw, and What It Did (Norwich, NY: J. F. Hubbard Jr., 1866), 152; [John Fisk Allen], *Memorial of Pickering Dodge Allen* (Boston: Henry W. Dutton and Son, 1867), 112.

The destruction of the *Diana* is examined in Horace J. Beach, "The Last Moments of the Gunboat *Diana*, and Her Almost Final Resting Place," 2010, 49 pp. (PDF document), The Young-Sanders Center for the Study of the War between the States in Louisiana, http://www.youngsanders.org/Thearticle.pdf.

14. Frazier, *Thunder across the Swamp*, 330–31, 335–36, 345; De Forest, "Forced Marches," 710; Hosmer, *The Color-Guard*, 137; Browning, *Lincoln's Trident*, 305.

15. Frazier, *Thunder across the Swamp*, 345, 349–50, 351.

16. Ibid., 349–51; *OR*, Series I, Vol. 15, 343–44; Raphael, *Battle in the Bayou Country*, 117.

17. *Official Report Relative to the Conduct of Federal Troops in Western Louisiana, during the Invasions of 1863 and 1864* (Shreveport: John Dickinson, 1865), 39.

The families of Stirling, Wilcoxson, Fuselier, Charpentier, Cornay, Perkins, Bethel, Harding, and Byrne all owned sugar plantations along the Teche. See P. A. Champomier, *Statement of the Sugar Crop of Louisiana of 1861–62* (New Orleans: Cook, Young, 1862), 31–33.

David C. Edmonds examined the second invasion of the Teche in *Yankee Autumn in Acadiana: A Narrative of the Great Texas Overland Expedition through Southwestern Louisiana, October-December 1863* (Lafayette, LA: Acadiana Press, 1979).

No historian has yet written a detailed study of the third invasion of the Teche, but cursory treatments can be found in Gary Dillard Joiner, *One Damn Blunder from Beginning to End: The Red River Campaign of 1864* (Wilmington, DE: Scholarly Resources, 2003), 55, 57; Gary D. Joiner, *Through the Howling Wilderness: The 1864 Red River Campaign and Union Failure in the West* (Knoxville: University of Tennessee Press, 2006), 63, 64; Michael J. Forsyth, *The Red River Campaign of 1864 and the Loss by the Confederacy of the Civil War* (Jefferson, NC: McFarland, 2002), 53–54.

18. *OR*, Series I, Vol. 15, 373; Frazier, *Thunder across the Swamp*, 524–26, 531–41; Raphael, *Battle in the Bayou Country*, 137, 162–66; John G. Mudge, "A Night on Picket," *Stories of Our Soldiers: War Reminiscences* (Boston: Journal Newspaper Company, 1893), 181; "The Passing of Harry Mahoney, Negro Politician of Radical Days," typewritten transcript (Summer 1956) of article from unidentified New Orleans newspaper [ca. 1903–10?], Gray Osborn Collection, Avery Island, Inc., Archives, Avery Island, LA.

Chapter 5

1. Conrad, "New Iberia and Yellow Fever," 174; "Louisiana Intelligence," *New Orleans Daily Crescent*, February 19, 1866, 2.

2. Statistics for antebellum sugar production in St. Mary Parish are found in Richard Follett, *Documenting Louisiana Sugar, 1845–1917* (Swindon: Arts & Humanities Research Council [UK], 2008), Internet database, www.sussex.ac.uk/louisianasugar/, accessed

September 14, 2014; statistics for postbellum sugar production in St. Mary, St. Martin, and Iberia parishes are found in L. Bouchereau, *Statement of the Sugar and Rice Crops Made in Louisiana* (1868–1869), 48; (1869–1870), 69, 72, 75; (1870–1871), 66; (1872–1873), 80; (1873–1874), 72, 75, 77.

As early as 1864 French diplomat Charles Prosper Fauconnet discussed the challenges of sugar planting versus the benefits of cotton planting in Louisiana; see Carl A. Brasseaux and Katherine Carmines Mooney, eds., *Ruined by This Miserable War: The Dispatches of Charles Prosper Fauconnet, a French Diplomat in New Orleans, 1863–1868* (Knoxville: University of Tennessee Press, 2012), 65. Carl A. Brasseaux, Lafayette, LA, e-mail to the author, September 18, 2014.

3. The price of middling cotton per pound, New York market, in 1866 was $43.20; in 1870, $23.98; in 1876, $12.98. The wholesale price of refined, crushed, and granulated sugar per pound in April 1866 was $16; in April 1872, $12; in January 1878, $8.62; in 1910 (wholesale granulated), $4.97.

Cotton prices, 1866–1876, in "Average Prices of Middling Cotton," *Statistical Abstract of the United States, 1898* (Washington: U.S. Government Printing Office, 1899), 394; Sugar prices, 1866–1878, in "Sugar, Refined, Crushed, and Granulated," *Movement of Prices, 1840–1899* (Washington: Bureau of Statistics, Treasury Department / [U.S. Government Printing Office, 1900]), 3172; Sugar price, 1910, in "Sugar Prices," *Hearings Held before the Special Committee on the Investigation of the American Sugar Refining Co. and Others, House of Representatives, Vol. 2* (Washington: U.S. Government Printing Office, 1911), 1493.

"St. Mary," *Louisiana Planter and Sugar Manufacturer* (February 13, 1904): 112; "Cane Versus Cotton on the Teche," *Louisiana Planter and Sugar Manufacturer* (February 26, 1898): 131.

4. Captain E. B. Trinidad, *Rough Sketch* [of the Teche], map (n.p., ca. 1868), Cartographic Division, National Archives and Records Administration, Washington [reproduction obtained by author from Grevemberg House Museum, Franklin, LA]; Brasseaux, *Acadian to Cajun*, 75, 83, 86–87.

Carl A. Brasseaux's discussion of postbellum economic strains on planters of Acadian descent applies to other planters along the Teche, regardless of ethnicity. See Brasseaux, *Acadian to Cajun*, 74–88.

Additional references to northerners buying up plantations along the postbellum Teche appear in *Report of the Commissioner of Agriculture for the Year 1877* (Washington: U.S. Government Printing Office, 1878), 29; McKinstry, "The Bayou Teche," 168.

5. "Louisiana Sugar News," *Louisiana Planter and Sugar Manufacturer* (May 29, 1920): 346.

6. L. Bouchereau, *Statement of the Sugar and Rice Crops* (1868–69), 48; A. Bouchereau, *Statement of the Sugar and Rice Crops Made in Louisiana* (1885–86), 53; Charles Tenney Jackson, *The Fountain of Youth* (New York: Outing Publishing, 1914), 161; *Encyclopedia of Tariffs and Trade in U.S. History*, s.v. "Sugar"; Edwin J. Foscue and Elizabeth Troth, "Sugar Plantations of the Irish Bend District, Louisiana," *Economic Geography* 12 (October 1936): 376–77, 380; "Raw Sugar Factories" (including co-ops), American Sugar Cane League, 2015, www.amscl.org/factories, accessed June 7, 2015.

7. Eric Foner, *Reconstruction: America's Unfinished Revolution, 1863–1877* (New York: Perennial, 1988), 119–23; Michael S. Martin, "Violent Louisiana: Chaos after the Civil War," public lecture, Jeanerette Museum, Jeanerette, LA, August 21, 2013, audio recording in possession of the author; John C. Rodrigue, *Reconstruction in the Cane Fields: From Slavery to Free Labor in Louisiana's Sugar Parishes, 1862–1880* (Baton Rouge: Louisiana State University Press, 2001), 98–102, 183–88; Raymond D. Gastil, "Violence, Crime, and Punishment," *Encyclopedia of Southern Culture* (Chapel Hill: University of North Carolina Press, 1989), 1475.

8. Rodrigue, *Reconstruction in the Cane Fields*, 100, 166–67; Brasseaux, *Acadian to Cajun*, 137–38, 143–44; James G. Dauphine, "The Knights of the White Camellia and the Election of 1868: Louisiana's White Terrorists; a Benighting Legacy," *Louisiana History* XXX (Spring 1989): 179.

9. *Supplemental Report of Joint Committee of the General Assembly of Louisiana on the Conduct of the Late Elections* (New Orleans: A. L. Lee, 1869), 137–38. Although the original document runs these incidents together in a single paragraph, I have given each a paragraph of its own for ease of reading.

10. Brasseaux, *Acadian to Cajun*, 144–46; George C. Rable, *But There Was No Peace: The Role of Violence in the Politics of Reconstruction* (Athens: University of Georgia Press, 2007), 130; "The Trouble in the State of Louisiana," (Alexandria) *Louisiana Democrat*, June 11, 1873, 3; "Use of the Army in Certain of the Southern States," Executive Document No. 30 (January 24, 1877), in *Index to the Executive Documents of the House of Representatives for the Second Session of the Forty-Fourth Congress, 1876-'77, Vol. 9* (Washington: U.S. Government Printing Office, 1877), 222; "Condition of the South," Report No. 261 (23 February 1875), in *Index to Reports of Committees of the House of Representatives for the Second Session of the Forty-Third Congress, 1875-'75, Vol. 5* (Washington: U.S. Government Printing Office, 1875), 778, 779.

11. "Lynching of a Brutal Negro," *New York Times*, December 4, 1878, 1; Michael J. Pfeifer, "Lynching and Criminal Justice in South Louisiana, 1878–1930," *Louisiana History* XL (Spring 1999): 164; "The Lynchers, Inflict Summary Vengeance on a Negro at New Iberia," *Daily Picayune*, January 31, 1889, 1; Gilles Vandal, "Politics and Violence in Bourbon Louisiana: The Loreauville Riot of 1884 as a Case Study," *Louisiana History* XXX (Winter 1989): 23, 38; Gilles Vandal, *Rethinking Southern Violence: Homicides in Post-Civil War Louisiana, 1866–1884* (Columbus: Ohio State University Press, 2000), 201; Herbert Shapiro, *White Violence and Black Response: From Reconstruction to Montgomery* (Amherst: University of Massachusetts Press, 1988), 27; Rodrigue, *Reconstruction in the Cane Fields*, 186.

12. John M. Woodworth, M.D., *The Cholera Epidemic of 1873 in the United States* (Washington: U.S. Government Printing Office, 1875), 118–19; Conrad, "New Iberia and Yellow Fever," 175–78, 181, 186–87; Carrigan, *The Saffron Scourge*, 4, 101–03, 191, 166, 194; Shakspeare Allen, M.D., "Remarks on the Infection of Yellow Fever and Its Portability," *Annual Report of the Board of Health, to the General Assembly of Louisiana, December 31st, 1871, Session of 1872* (New Orleans: n.p., 1872), 68; *New Orleans Republican*, August 31, 1867, 2; (Franklin, LA) *Planters' Banner*, December 28, 1867, 2.

13. Conrad, "New Iberia and Yellow Fever," 174–75; William F. Switzler, "The Levees," *Report on the Internal Commerce of the United States, Commerce and Navigation, Pt. II* (Washington: U.S. Government Printing Office, 1888), 246–48; "Cane Sugar as an Article of American Industry," *Agricultural Review and Industrial Monthly* (February 1884): 163; "The Teche, an Alarming State of Affairs on the Bayou," (New Orleans) *Daily Picayune*, April 2, 1882, 1; "The Teche Country, the Water Rising an Inch an Hour," (New Orleans) *Daily Picayune*, April 2, 1882, 1; "The Teche Country, a Diminished Rise in the Past Day," (New Orleans) *Daily Picayune*, April 6, 1882, 1; "Relics of the Inundation," (New Orleans) *Daily Picayune*, May 9, 1882, 6.

14. John M. Barry, *Rising Tide: The Great Mississippi Flood of 1927 and How It Changed America* (New York: Touchstone, 1998), 173–76; Glenn R. Conrad and Carl A. Brasseaux, *Crevasse: The 1927 Flood in Acadiana* (Lafayette: Center for Louisiana Studies, University of Southwestern Louisiana, 1994), 21–24, 31.

15. Conrad and Brasseaux, *Crevasse*, 24, 26–29; Glenn R. Conrad, "The Teche Country in the Flood of 1927," in Glenn R. Conrad, ed., *New Iberia: Essays on the Town and Its People*, 2nd ed. (Lafayette: Center for Louisiana Studies, University of Southwestern Louisiana, 1986), 357, 359; "Residents Warned to Flee 27 Towns in Path of Flood," (New Orleans) *Times-Picayune*, May 19, 1927, 1.

16. Conrad and Brasseaux, *Crevasse*, 27, 29, see also "before" and "after" images, 97, 98; George Healy, "Waters Invading St. Martin Parish," (New Orleans) *Times-Picayune*, May 20, 1927, 1; George Healy, "Panic-Stricken Thousands Flee before Flood in Teche," (New Orleans) *Times-Picayune*, May 21, 1927, 3.

17. Conrad and Brasseaux, *Crevasse*, 31–32, 45; Healy, "Panic-Stricken Thousands," 3; Isaac Monroe Cline, *Storms, Floods and Sunshine*, 3rd ed. (New Orleans: Pelican, 1951), 204.

18. Mike Dunne, "A Flood of Memories: The Great Flood of 1927 Changed Lives and History in Louisiana," (Baton Rouge) *Advocate*, May 12, 2002, accessed through the Advocate Archives, www.theadvocate.com/archives/, accessed April 22, 2015; George Healy, "Panic-Stricken Thousands," 3; "Torrent Raging in Teche Moves on New Iberia," (New Orleans) *Times-Picayune*, May 22, 1927, 1; George Healy, "Water Climbing Fast on Ridge of Bayou Teche," (New Orleans) *Times-Picayune*, May 25, 1927, 2.

19. Conrad and Brasseaux, *Crevasse*, 27, 29, 44–45; Cline, *Storms, Floods and Sunshine*, 210.

20. Conrad and Brasseaux, *Crevasse*, 29–30, 65–66; Conrad, "Teche Country in the Flood of 1927," 362; *Mississippi River Flood of 1927 Showing Flooded Areas and Field of Operations*, map (Washington: U.S. Coast and Geodetic Survey/U.S. Army Corps of Engineers, 1927); Flood of 1927 high-water marks on Lutzenberger Foundry, New Iberia, LA, measured by author with assistance from building owner Reving Broussard, December 15, 2014.

21. Conrad and Brasseaux, *Crevasse*, 1, 39 (Table I), 41 (Table II), 45–72, 75. Conrad and Brasseaux included no data for St. Mary Parish in the above two tables in *Crevasse*. A detailed 1927 flood map, however, shows about 90 percent of St. Mary Parish underwater. In addition, the author estimated the number of St. Mary Parish flood victims by using 1930 U.S. Census data. Specifically, by adding the population of Morgan City (most of

which flooded) to the population of the parish's outlying areas (almost all of which flooded), while excluding the populations of Franklin, Berwick, and Patterson (which either did not flood or experienced relatively minor flooding). The estimated number of St. Mary Parish flood victims thus totaled approximately 22,241. See *Mississippi River Flood of 1927*, map; *Fifteenth Census of the United States, 1930, Distribution, Vol. I, Retail Distribution, Pt. I, Summary for the United States, and Statistics for Counties and Incorporated Places of 1,000 Population and Over*, "Table 13, Louisiana, Retail Distribution, by Parishes and Incorporated Places [includes parish population]," 137.

Chapter 6

1. Captain E. B. Trinidad, *Rough Sketch* [of the Teche], map (n.p., ca. 1868), Cartographic Division, National Archives and Records Administration, Washington, DC [reproduction obtained by author from Grevemberg House Museum, Franklin, LA].

I have drawn the title of this chapter from Martin Reuss's excellent *Designing the Bayous: The Control of Water in the Atchafalaya Basin, 1800–1995* (College Station: Texas A&M University Press, 2004).

2. "Report," *Report of the Board of Public Works to the General Assembly of Louisiana, for the Year 1870* (New Orleans: *The Republican*, 1871), 4; Report of C. W. Howell, Captain [of] Engineers and Brevet Major U.S.A., *Annual Report of the Chief of Engineers to the Secretary of War for the Year 1870* (Washington: U.S. Government Printing Office, 1870), 347–51; Captain E. B. Trinidad, New Orleans, to Major C. W. Howell, Captain of Engineers, New Orleans, June 12, 1872, *Annual Report of the Chief of Engineers to the Secretary of War for the Year 1872* (Washington: U.S. Government Printing Office, 1872), 568; Routh Trowbridge Wilby, *Clearing Bayou Teche after the Civil War: The Kingsbury Project, 1870–1871* (Lafayette: Center for Louisiana Studies, University of Southwestern Louisiana, 1991), x, 1; Brasseaux and Fontenot, *Steamboats on Louisiana's Bayous*, 116–17.

3. Wilby, *Clearing Bayou Teche*, 1, 6, 7.

4. Ibid., 65, 67, 84, 86, 87, 89, 92, 94, 102.

5. Ibid., 77, 78, 82, 103; Trinidad to Howell, 568.

6. "Improvement of Harbor at New Orleans; of Pearl River, Mississippi; of Bayous Teche and Black, and Other Water-Courses in Louisiana—Connection of Bayou Teche with Grand Lake at Charenton," *Annual Report of the Chief of Engineers, United States Army, to the Secretary of War, for the Year 1881, Pt. I* (Washington: U.S. Government Printing Office, 1881), 194; "Probable Water-Supply, Location and Estimated Cost of Building a Lock and Dam for Making Slackwater Navigation on the Upper Portion of Bayou Teche, Louisiana," *Annual Report of the Chief of Engineers, United States Army, to the Secretary of War, for the Year 1889, Pt. II* (Washington: U.S. Government Printing Office, 1889), 1516; "Improvement of Bayou Teche," 1159.

7. "Report of Lieutenant O. T. Crosby, Corps of Engineers," *Annual Report of the Chief of Engineers, United States Army, to the Secretary of War, for the Year 1887, Pt. II* (Washington: U.S.

Government Printing Office, 1887), 1371; "Report of Mr. H. C. Collins, Assistant Engineer," January 31, 1880, *Annual Report of the Chief of Engineers, United States Army, to the Secretary of War[,] for the Year 1880, Pt. II* (Washington: U.S. Government Printing Office, 1880), 1167.

8. "Improvement of Harbor at New Orleans; of Pearl River, Mississippi; of Bayous Teche and Black," 194; "Letter from the Secretary of War Transmitting a Report from Maj. Amos Stickney, Corps of Engineers, As Containing the Information Called for by Senate Resolution of March 15, 1882, Relative to Bayous Courtableau, Teche, and Terrebonne, Louisiana," Executive Document No. 146 (March 31, 1882), in *Executive Documents Printed by Order of the Senate of the United States for the First Session of the Forty-Seventh Congress, Vol. V* (Washington: U.S. Government Printing Office, 1882), 2; "Report of Mr. H. C. Collins, Assistant Engineer," June 30, 1883, *Annual Report of the Chief of Engineers, United States Army, to the Secretary of War, for the Year 1883, Pt. II* (Washington: U.S. Government Printing Office, 1883), 1116.

9. "Letter from the Secretary of War Transmitting a Report from Maj. Amos Stickney," 2; "Report of Lieutenant O. T. Crosby," 1371.

10. Brasseaux and Fontenot, "Number of Bayou Country Steamboats Docking at New Orleans, 1845–1860," Table 3.2, *Steamboats on Louisiana's Bayous*, 50; William F. Switzler, "Steam-boat Arrivals at the Port of New Orleans for the Year Ending August 31, 1880" and "Steam-boats Running to New Orleans in 1886," *Report on the Internal Commerce of the United States, Pt. II: Commerce and Navigation* (Washington: U.S. Government Printing Office, 1888), 137–38, 230. Fontenot and Brasseaux offer data on the number of steamboat arrivals at New Orleans from Bayou Teche for the years 1845, 1850, 1855, and 1860; Switzler offers data on the same for 1886 and the year ending August 31, 1880.

11. Brasseaux and Fontenot, *Steamboats on Louisiana's Bayous*, 126–27; Donald J. Millet, "Southwest Louisiana Enters the Railroad Age: 1880–1900," *Louisiana History* XXIV (Spring 1983): 169, 173–74; Jim Bradshaw, *100 Years on the River: The Chotin Family and Their Boats* (Lafayette: Center for Louisiana Studies, University of Louisiana at Lafayette, 2001), 34–35; *Map of Louisiana* (Chicago: Rand McNally, 1901), Historical Map Archive, University of Alabama, alabamamaps.ua.edu/historicalmaps/, accessed November 1, 2014.

12. "Report of Lieutenant O. T. Crosby, Corps of Engineers," *Annual Report of the Chief of Engineers, United States Army, to the Secretary of War, for the Year 1885, Pt. II* (Washington: U.S. Government Printing Office, 1885), 1439; "Survey of Atchafalaya River, Louisiana, above Berwick Bay," *Annual Report of the Chief of Engineers, United States Army, to the Secretary of War, for the Year 1885, Pt. II* (Washington: U.S. Government Printing Office, 1885), 1434; Brasseaux and Fontenot, *Steamboats on Louisiana's Bayous*, 124, 125, 128–29; W. F. Allen, comp. and ed., "Morgan's Louisiana and Texas Railroad," *Travelers' Official Railway Guide for the United States and Canada* (July 1882): 313.

13. Brasseaux and Fontenot, *Steamboats on Louisiana's Bayous*, 130–35; "*Houston v. The Police Jury of St. Martin [Parish]*," in Merritt M. Robinson, *Reports of Cases Argued and Determined in the Supreme Court of Louisiana for the Years 1848, Vol. III* (New Orleans: T. Rea, 1849), 566–67; Philip Graham, *Showboat: The History of an American Institution* (Austin: University of Texas Press, 1969), 63; "Report of Major W. H. Heuer," 1248.

I speculate that the troupe brought to Arnaudville aboard the *Hattie Bliss* in spring 1886 was DeVere's because that same spring the *Hattie Bliss* brought DeVere's troupe to Donaldsonville, Louisiana. Untitled article about DeVere's Carnival of Novelties and the Rex Silver Cornet Band, *Donaldsonville Chief*, March 20, 1886, 2; untitled article about DeVere's Carnival of Novelties and Mastodon Dog Circus, *Donaldsonville Chief*, March 27, 1886, 3; untitled article referring to *Hattie Bliss* as carrying DeVere's Carnival of Novelties, *Donaldsonville Chief*, April 3, 1886, 3; untitled article about DeVere's Carnival of Novelties, *Donaldsonville Chief*, April 10, 1886, 2.

14. Brasseaux and Fontenot, *Steamboats on Louisiana's Bayous*, 135–36; "Improvement of Bayou Teche, from Saint Martinsville to Port Barre, Louisiana," *Annual Report of the Chief of Engineers, United States Army, to the Secretary of War for the Year 1880, Pt. II* (Washington: U.S. Government Printing Office, 1880), 1159.

The post-1909 steamboat travel time between New Iberia and New Orleans is mentioned in "New Boat to Run into Bayou Teche," *St. Mary Banner*, August 13, 1921, 6; compare to the pre-1909 travel time mentioned in Brasseaux and Fontenot, *Steamboats on Louisiana's Bayou*, 125.

15. "Building a Lock and Dam for Making Slackwater Navigation on the Upper Portion of Bayou Teche," 1516; "Southwest Louisiana's Biggest Project," 1; "Fix Hearing on Cocodrie Closing for Next Tuesday," (Opelousas, LA) *St. Landry Clarion*, January 22, 1921, 1; Jim Bradshaw, "Keeping Water in the Teche a Priority of the 20th Century," TecheToday.com, July 29, 2009, www.techetoday.com/keeping-water-teche-priority-20th-century, accessed August 26, 2014.

Although work on the projects would continue into 1921, the excavation of Ruth Canal and the dredging of the upper Teche permitted direct navigation from Port Barre to Lafayette (per Ruth Canal) by September 1920. See "Southwest Louisiana's Biggest Project," 1; "Brown Says Dredge Is Due to Begin Next Week," (Opelousas, LA) *St. Landry Clarion*, July 2, 1921, 1; "Federal Engineer Inspects the Newly Dredged Teche," (Opelousas, LA) *St. Landry Clarion*, August 13, 1921, 8.

16. "Improvement of Rivers and Harbors in the New Orleans, Louisiana, District," *Report of the Chief of Engineers, U.S. Army, 1909, Pt. I* (Washington: U.S. Government Printing Office, 1909), 449–50; "Improvement of Rivers and Harbors in the New Orleans, Louisiana, District," *War Department, Annual Reports, 1910, Vol. II, Report of the Chief of Engineers* (Washington: U.S. Government Printing Office, 1910), 502; "Improvement of Rivers and Harbors in the New Orleans, Louisiana, District," *Report of the Chief of Engineers, U.S. Army, 1912, Pt. I* (Washington: U.S. Government Printing Office, 1912), 656–57; "Abstract of Contracts in Force during Fiscal Year Ending June 30, 1912," *Report of the Chief of Engineers, U.S. Army, 1913, Pt. I* (Washington: U.S. Government Printing Office, 1912), 1967–69; "Improvement of Rivers and Harbors in the New Orleans, La., District," *Report of the Chief of Engineers, U.S. Army, 1913, Pt. 1* (Washington: U.S. Government Printing Office, 1913), 721–23; "Improvement of Rivers and Harbors in the New Orleans, La., District," *Report of the Chief of Engineers, U.S. Army, 1915, Pt. I* (Washington: U.S. Government Printing Office, 1915), 834, 836.

The superintendent of Keystone Plantation, John Peters, stated in 1899 that the canal running from Spanish Lake to the plantation existed as early as around 1845. A state plat map dated 1845 appears to show the canal, identified as "mill race." "Mr. John Peters Disclaims Lowering the Level of Spanish Lake," 2; Plat map for T11S, R6E, Sec. 17, Southwest District, approved December 23, 1845, Office of State Lands, State of Louisiana, Baton Rouge, LA.

17. Conrad, "Road for Attakapas," 32–36; "The Old Spanish Trail Highway," map, (New Orleans) *Times-Picayune*, October 15, 1916, C-13; "Touring Information for Motorists," (New Orleans) *Times-Picayune*, March 7, 1920, Sec. 3, 15; Vic Calver, "Louisiana's Main Trunk Highway System[,] Seventeen Hundred Miles in Length, Will Link Parish Seats with Nation's Roads," (New Orleans) *Times-Picayune*, March 21, 1920, Sec. 3, 1; "Beauties of Spanish Trail Told in Bureau's Article," (New Orleans) *Times-Picayune*, September 15, 1929, Sec. 3, 5. The overlap of the Old Spanish Trail and the Pershing Highway along Bayou Teche is documented in "Auto Trails Map," *Commercial Atlas of America* (Chicago: Rand McNally, 1924), 408–9; see also David Rumsey Map Collection, www.davidrumsey.com/luna/servlet/detail/RUMSEY~8~1~201735~3000680, accessed November 9, 2015.

18. "Map of Louisiana Showing Progress of Hardsurfacing Program," in Lemmon et al., *Charting Louisiana*, 273 (map 150); *Louisiana Official Road Map, Winter Edition 1937–1938* ([Baton Rouge]: Louisiana Highway Commission, 1937–38).

19. "New Orleans, La., District," *Annual Report of the Chief of Engineers, 1924, Extract: Commercial Statistics for the Calendar Year 1923* (Washington: U.S. Government Printing Office, 1924), 865; "Report of Major H. M. Adams, Corps of Engineers, Officer in Charge, for the Fiscal Year June 30, 1900, with Other Documents Relating to the Work," *Annual Reports of the War Department for the Fiscal Year Ended June 30, 1900, Report of the Chief of Engineers, Pt. 3* (Washington: U.S. Government Printing Office, 1900), 2261; "New Orleans, La., District," *Report of the Chief of Engineers, U.S. Army, 1933, Pt. 2, Commercial Statistics, Water-Borne Commerce of the United States for the Calendar Year 1932* (Washington: U.S. Government Printing Office, 1933), 556; Brasseaux and Fontenot, *Steamboats on Louisiana's Bayous*, 129, 135; Graham, *Showboat*, 185–87.

Passenger statistics are based on the average number of passengers during the years 1924, 1925, 1928, 1929, 1930, and 1931.

20. [Benjamin Moore Norman], *Norman's New Orleans and Environs* (New Orleans: B. M. Norman, 1845), 33; Richardson, "The Teche Country Fifty Years Ago," 122; James P. Kemper, "Down Where the Sugar Cane Grows: The Reminiscences of James P. Kemper, Part I: It Was Nature's Idea," *Attakapas Gazette* XV (Winter 1980): 158.

21. "St. Mary (Special Correspondence)," *Louisiana Planter and Sugar Manufacturer*, August 5, 1899, 87; Benjamin D. Maygarden and Jill-Karen Yakubik, *Bayou Chene: The Life Story of an Atchafalaya Basin Community* ([New Orleans]: U.S. Army Corps of Engineers, 1999), 12–13; "Letter from the Secretary of War, Transmitting, with a Letter from the Chief of Engineers, Report of Preliminary Examination and Survey of Bayou Teche, La., with View to Securing Increased Depth," Document No. 1329, *House Documents, Vol. 28, 62nd Congress, 3rd Sess.* (December 2, 1912–March 4, 1913) (Washington: U.S.

Government Printing Office, 1913), 7; Jno. E. Williams, "Fragments, Perhaps a Serial, of Cypress History," *St. Louis Lumberman* (April 15, 1917), 38; Jno. E. Williams, "Fragments of Cypress History and Comments—Part II," *St. Louis Lumberman* (May 1, 1917), 42–43; *National Lumber Manufacturers' Association, Official Report, Eighth Annual Convention, Held in New Orleans, Louisiana, April 19 and 20, 1910* (Chicago: National Lumber Manufacturers' Association, 1910), 253–54; "R. H. Downman and Louisiana Red Cypress," *Lumber Trade Journal* (August 1, 1901), 30–31. See also Blair Bonin et al., *Where the Bayou Runs Straight: The History of Jeanerette* (Jeanerette, LA: N.p., 1982), 228–29; Donna Fricker, "The Louisiana Lumber Boom, c. 1880–1925" ([Baton Rouge]: Fricker Historic Preservation Services, [2012?]), 1–2; Brasseaux and Fontenot, *Steamboats on Louisiana's Bayous*, 138; "Why Louisiana Conserves Her Forests," *Proceedings of the Louisiana Engineering Society* VII (1921): 121.

The passage quoted from "Why Louisiana Conserves Her Forests" actually reads "*without* little benefit to anyone"; it is clear from context, however, that the author meant "*with* little benefit to anyone," and I have corrected the text accordingly.

Although Maygarden and Yakubik examined the lumber industry on Bayou Chêne, their general observations apply equally well to the industry on Bayou Teche.

22. Brasseaux and Fontenot, *Steamboats on Louisiana's Bayous*, 138; Bradshaw, *100 Years on the River*, 50–51, see also Appendix 3, s.v. *Albert Hanson*, *Amy Hewes*, *M. E. Norman*; Jim Bradshaw, "Last Paddle-Wheeler on the Teche Brought Groceries," TecheToday.com, March 8, 2012, www.techetoday.com/last-paddle-wheeler-teche-brought-groceries, accessed November 16, 2014; Wilda B. Moran, "Bringing in Supplies by Steamboat," (New Iberia, LA) *Daily Iberian*, Bicentennial Special Edition, clipping in the vertical files of the Iberia Parish Library, New Iberia, LA, from historical database in possession of Carl A. Brasseaux, Lafayette, LA; "Up and Down the Street," (New Orleans) *Times-Picayune*, March 23, 1935, 28.

23. "1928 Flood Control Act," 70th Cong., 1st Sess. (1928), U.S. Army Corps of Engineers, www.mvd.usace.army.mil/Portals/52/docs/MRC/Appendix_E._1928_Flood_Control_Act.pdf, accessed December 7, 2014; "West Atchafalaya Basin Protection Levee," *Waterbody Management Plan Series: Atchafalaya Basin, Lake History & Management Issues*, Pt. VI-A, Inland Fisheries Section, Office of Fisheries, Louisiana Department of Wildlife and Fisheries ([Baton Rouge]: Louisiana Department of Wildlife and Fisheries, 2009; updated 2014), 23; "Mississippi River Commission," *Annual Report of the Chief of Engineers, United States Army, 1930, [Pt. 1?]* ([Washington: U.S. Government Printing Office, 1930]), 2081; "Mississippi River Commission," *Report of the Chief of Engineers, U.S. Army, 1942, Pt. 1, Vol. 2* (Washington: U.S. Government Printing Office, 1943), 1999.

24. Reuss, *Designing the Bayous*, 156; Whitney P. Broussard III, Bayou Teche vector diagram, in "Water Quality Monitoring in the Bayou Teche Watershed, Final Report" (Lafayette: Institute for Coastal Ecology and Engineering, University of Louisiana at Lafayette, 2013), 8.

25. Conrad and Brasseaux, *Crevasse*, 19; "Public Hearings on Bridges over Channel Called," (New Orleans) *Times-Picayune*, April 10, 1938, Sec. 5, 8; Broussard, Bayou Teche

vector diagram; "Mississippi River Commission," *Report of the Chief of Engineers, U.S. Army, 1940, Pt. 1, Vol. 2* (Washington: U.S. Government Printing Office, 1941), 2275.

26. "Bids to Excavate Channel Sought," (New Orleans) *Times-Picayune*, August 29, 1937, 6; "Award Approved for Atchafalaya Wax Lake Outlet," (New Orleans) *Times-Picayune*, October 21, 1937, 8; "Public Hearings on Bridges over Channel Called," Sec. 5, 8; "Construction of Levees to Protect Bayou Teche Slated to Start Soon," (New Orleans) *Times-Picayune*, October 17, 1948, 20; "Bayou Teche Project Bids to Open June 7," (New Orleans) *Times-Picayune*, May 26, 1950, 46; Broussard, Bayou Teche vector diagram.

Aerial photographs overlaid digitally with military maps from the Civil War indicate that Wax Lake Outlet ran directly through the Fort Bisland site.

27. Brasseaux and Fontenot, *Steamboats on Louisiana's Bayous*, 67, 137, 149–51; Bradshaw, *100 Years on the River*, 102; Lynn M. Alperin, *History of the Gulf Intracoastal Waterway*, Navigation History NWS-83-9 ([Alexandria, VA]: National Waterways Study, U.S. Army Engineer Water Resources Support Center, Institute for Water Resources, January 1983), 36; Adam I. Kane, *The Western River Steamboat* (College Station: Texas A&M University Press, 2004), 16 (Table 1.1); Shane K. Bernard, Bayou Teche shipping database, 1891–1997, in possession of the author, New Iberia, LA; *Waterborne Commerce of the United States, Calendar Year 1997, Pt. 2, Waterways and Harbors: Gulf Coast, Mississippi River System and Antilles* (Fort Belvoir, VA: Water Resources Support Center, U.S. Army Corps of Engineers, [1997?]), 185–86.

28. William Langewiesche, "Down on the Bayou," *Flying Magazine* (May 1979): 124.

Conclusion

1. "About Us," The TECHE Project, www.techeproject.org/about/, accessed December 31, 2014; Dane Thibodeaux, n.p., Bayou Operations Coordinator, the TECHE Project, e-mail to the author, December 27, 2014. The word *TECHE* in "the TECHE Project" is an acronym meaning "Teche Ecology, Culture and History Education." The author is a member of the TECHE Project.

2. *Bayou Têche Paddle Trail Planning 2012–2017*, September 2012, www.techeproject.org/wp-content/uploads/2013/05/BTPT-plan-final.pdf, accessed December 31, 2014, 4, 11; Bayou Teche Paddle Trail, fold-out map (Arnaudville, LA: The TECHE Project, 2014); Annie Ourso, "Bayou Teche Canoe, Kayak Trail Plans Moving Forward," *(Baton Rouge) Advocate*, January 10, 2014, www.theadvocate.com/home/7898865-125/bayou-teche-canoe-kayak-trail, accessed December 31, 2014; "New Water Trails Designated in Louisiana and Michigan," press release, National Park Service, January 21, 2015, www.nps.gov/news/release.htm?id=1671, accessed January 27, 2015; Dominick Cross, "Bayou Teche Paddle Trail Nets National Designation," theadvertiser.com, January 21, 2015, www.theadvertiser.com/story/news/local/2015/01/21/bayou-teche-paddle-trail-nets-national-designation/22119869/, accessed January 27, 2015; William Johnson, "Bayou Teche Added to National Water Trails System," (Opelousas, LA) *Daily World*, January 21, 2015, www.dailyworld.com/story/news/

local/2015/01/21/bayou-teche-added-national-water-trails-system/22126727/, accessed January 15, 2015.

3. Blake Couvillion, n.p., founder, Cajuns for Bayou Teche, e-mail to the author, December 27, 2014; Kristen Kordecki, n.p., co-founder and former executive director, the TECHE Project, e-mail to the author, December 29, 2014; "Bring Back the Paddle," *Acadiana Gazette*, June 4, 2014, 2-B, www.acadianagazette.net/archives/volume10/issue23/2014_6_04_p8.pdf, accessed December 31, 2014; "Cajuns for Bayou Teche Founder Applauds Universal Garbage Pickup for St. Martin," techetoday.com, March 21, 2012, www.techetoday.com/cajuns-bayou-teche-founder-applauds-universal-garbage-pickup-st-martin, accessed December 31, 2014; "Dragons on Bayou Teche: Paddle, Watch, Help," techetoday.com, February 21, 2013, www.techetoday.com/dragons-bayou-teche-paddle-watch-help, accessed December 31, 2014.

4. "Main Race," Tour du Teche, www.tourduteche.com/index.php/fine-print/about, accessed December 31, 2014; "Tour du Teche in *Men's Journal*," Tour du Teche, November 2, 2014, www.tourduteche.com/index.php/74-tour-du-teche-in-men-s-journal, accessed December 31, 2014; Ken Grissom, "Tour du Teche Grows to 135 Miles, Four Parties," techetoday.com, August 11, 2011, www.techetoday.com/tour-du-teche-grows-135-miles-four-parties, accessed December 31, 2013; "Bring Back the Paddle," 2-B.

5. Hutchins, *An Historical Narrative*, 46; Mark A. Rees, ed., *Archaeology of Louisiana* (Baton Rouge: Louisiana State University Press, 2010), 23–24; Rees, "Plaquemine Mounds of the Western Atchafalaya Basin," 68; Weinstein, Kelley, and Saunders, *Expeditions of Clarence Bloomfield Moore*, 138–40.

6. Rees, "Plaquemine Mounds of the Western Atchafalaya Basin," 69, 70, 72, 78–87; Mark A. Rees, Lafayette, LA, e-mail to the author, December 28, 2014.

7. David Cheramie, "Looking for Beausoleil," *Acadiana Profile* (December 2013-January 2014), reprinted on myNewOrleans.com, ca. January 2014, www.myneworleans.com/Acadiana-Profile/December-January-2014/Looking-for-Beausoleil/, accessed December 31, 2014; Cain Burdeau, "Cajuns Seek Grave Site of Acadian Leader," *Bangor Daily News*, June 2, 2005, www.archive.bangordailynews.com/2005/06/02/cajuns-seek-grave-site-of-acadian-leader/, accessed January 1, 2005 [about Rees's field school of 2003]; Rees to author, December 28, 2014; "New Acadia Project Could Rewrite Area's History," Associated Press, September 2, 2013, n.p., reprinted on telegram.com, www.telegram.com/article/20130902/APN/309029932/0, accessed December 31, 2014.

8. For Teche-related archaeological studies by R. Christopher Goodwin & Associates, see R. Christopher Goodwin et al., *An Archeological and Historical Sites Inventory of Bayou Teche between Franklin and Jeanerette, Louisiana* (New Orleans: R. Christopher Goodwin & Associates, June 1985); R. Christopher Goodwin et al., *Historical and Archaeological Investigation of Fort Bisland and Lower Bayou Teche, St. Mary Parish, Louisiana* (New Orleans: R. Christopher Goodwin & Associates, June 1991); R. Christopher Goodwin, William P. Athens, and Allen R. Saltus Jr., *Evaluation of Magnetic Anomalies Located in Lower Bayou Teche, St. Mary Parish, Louisiana* (New Orleans: R. Christopher Goodwin & Associates, July 1991); R. Christopher Goodwin et al., *Supplemental Archaeological Investigation of*

Lower Bayou Teche, St. Mary Parish, Louisiana (New Orleans: R. Christopher Goodwin & Associates, August 1991); William P. Athens et al., *Phase I Cultural Resources Survey and Archaeological Inventory of the Proposed 19.3 KM (12 Mi) Long Stretch of Bayou Teche, Iberia Parish, Louisiana* (New Orleans: R. Christopher Goodwin & Associates, August 2000); William P. Athens et al., *Phase I Cultural Resources Survey and Archaeological Inventory of the 4.94 HA (12.21 AC) Keystone Lock and Dam Project Parcel, St. Martin Parish, Louisiana* (New Orleans: R. Christopher Goodwin & Associates, June 2001); William P. Athens et al., *Phase I Cultural Resources Survey and Archaeological Inventory of the Segura Staging Area, Iberia Parish, Louisiana* (New Orleans: R. Christopher Goodwin & Associates, April 2003).

9. See Jack R. Bergstresser Sr. et al., *National Register Evaluation of the Keystone Lock and Dam, St. Martin Parish, Louisiana* (Fort Walton Beach, FL: Prentice Thomas and Associates, December 1997); Ann E. Smith and Dave D. Davis, *Archaeological Excavations at Shadows-on-the-Teche* (New Orleans: Center for Archaeology, Tulane University, 1983); Hope Rurik, "Digging History: Professor, Students, Volunteers Seek Genesis of City," (New Iberia, LA) *Daily Iberian*, December 26, 2012, www.iberianet.com/news/digging-history-professor-students-volunteers-seek-genesis-of-city/article_ea0853b2–4f83–11e2-af5a-0019bb2963f4.html, accessed January 1, 2015; David T. Palmer, "Survey and Exploratory Excavation at the Lutzenberger Iron Foundry, New Iberia," *Newsletter of the Louisiana Archaeological Society* 40 (Winter 2012–13): 20; Charles E. Pearson, "Investigations of a Nineteenth Century Steamboat Wreck in Bayou Teche, Louisiana—A Possible Civil War Gunboat," *Louisiana Archaeology* 33 (2006 [published 2011]); Charles Larroque, "Guest Editorial," *kreole* (October-December 2013): 6; Jordan Kellman, "Seeking the Promised Land in Louisiana: The African Diaspora Heritage Trail Conference, 2013," *kreole* (October-December 2013): 31–33; Charles Larroque, Ray Brassieur, and Jihad Muhammad, prods., *The Creole Journey to the Promised Land* (documentary video, Lafayette, LA: Louisiane à la carte, 2013).

The place name *Promised Land* is rendered variously; alternate versions include Promiseland and Promise Land. I use the version that appears in Robert E. Maguire, "Hustling to Survive: Social and Economic Change in a South Louisiana Black Creole Community" (Ph.D. diss., McGill University, Montreal, Quebec, 1987), v, 18, passim.

The number of shipwrecks in Bayou Teche is cited in "Appendix I, Scope of Services," R. Christopher Goodwin et al., *Historical and Archaeological Investigation of Fort Bisland and Lower Bayou Teche, St. Mary Parish, Louisiana* (New Orleans: R. Christopher Goodwin & Associates, June 1991), 2.

10. *Draft Comprehensive Conservation Plan and Environmental Assessment: Bayou Teche National Wildlife Refuge, St. Mary Parish, Louisiana* (Atlanta: Fish and Wildlife Service [Southeast Region], U.S. Department of the Interior, May 2009), 13; "Bayou Teche National Wildlife Refuge," U.S. Fish and Wildlife Service, www.fws.gov/bayouteche/, accessed January 1, 2015; National Heritage Areas Act (creating Atchafalaya National Heritage Area), Public Law 109–338 (enacted October 12, 2006); "Atchafalaya National Heritage Area," www.atchafalaya.org, accessed January 1, 2015; "The Bayou Teche Wooden Boat Show," www.techeboatshow.com, accessed January 1, 2015; Bayou Teche Museum, www.bayoutechemuseum.org, accessed January 1, 2013; Holly Leleux-Thubron, "A Gift of the Past: Museum Offers

Timeline of City's History," (New Iberia, LA) *Daily Iberian*, December 27, 2010, www.iberianet.com/people/teche_life/a-gift-of-the-past/article_eb0ce376-57e4-5ad0-b36a-1475dce71c1f.html?mode=jqm, accessed January 1, 2015; Herman Fuselier, "Bayou Teche Event Celebrates Church, Acadians Arrival," (Lafayette, LA) *Daily Advertiser*, July 28, 2015, www.theadvertiser.com/story/entertainment/2015/07/28/bayou-teche-event-celebrates-church-acadians-arrival/30803181/, accessed August 18, 2015; "Eucharistic Procession down Bayou Teche," Diocese of Lafayette, Louisiana, ca. August 15, 2015, www.diolaf.org/events/eucharistic-procession-down-bayou-teche, accessed August 18, 2015.

11. *Bayou Têche Paddle Trail Planning 2012–2017*, 13; Garrie P. Landry, Department of Biology, University of Louisiana at Lafayette, e-mail to the author, January 6, 2015; "Experts Agree: Comprehensive Control Plan is Needed for Management and Eradication of Giant Salvinia," press release, House Committee on Natural Resources, U.S. Congress, June 27, 2011, www.naturalresources.house.gov/news/documentsingle.aspx?DocumentID=249072, accessed January 15, 2015.

12. Cinnamon Bair, "Water Hyacinth Was a Disaster," TheLedger.com, May 23, 2010, www.theledger.com/article/20100523/COLUMNISTS/5235028, accessed January 15, 2015; *Bayou Têche Paddle Trail Planning 2012–2017*, 13; "Cajuns for Bayou Teche to Remove Hyacinth in the Vermilion-Teche Watershed," KATC.com, June 28, 2012, www1.katc.com/news/cajuns-for-bayou-teche-to-remove-hyacinth-in-the-vermilion-teche-watershed/, accessed January 15, 2015; "War Declared on Water Lilies," techetoday.com, June 25, 2012, www.techetoday.com/war-declared-water-lilies, accessed January 15, 2015; "Removing Water Hyacinths from Louisiana Waters," *Annual Reports of the War Department for the Fiscal Year Ended June 30, 1900, Report of the Chief of Engineers, Pt. 3* (Washington: U.S. Government Printing Office, 1900), 2270–71; "Removing the Water Hyacinth, Louisiana," *Report of the Chief of Engineers, U.S. Army, 1920, Pt. 2* (Washington: U.S. Government Printing Office, 1920), 2410–11; "Removing the Water Hyacinth from Waters in Louisiana and Texas," *Report of the Chief of Engineers, U.S. Army, 1912, Pt. 1* (Washington: U.S. Government Printing Office, 1912), 680–81.

The role of sal soda in dispersing or dissolving arsenic for spraying purposes is noted in many ca. 1900 agricultural sources. See, for instance, L. R. Taft, "Spraying Calendar for 1898," *Michigan State Agricultural College Experiment Station Horticultural Department, Bulletin 155–156, March 1898* ([East Lansing]: Michigan State Agricultural College, 1898), 294.

Contrary to later claims, caustic soda does not appear to have been used in conjunction with arsenic to kill water hyacinths in Louisiana.

13. "Local News," (St. Martinville) *Weekly Messenger*, September 26, 1914, 3; Frederic J. Haskin, "Varieties in the Meat Supply," *Milwaukee Sentinel*, October 2, 1912, 6; "Removal of Water Hyacinth," *War Department Appropriation Bill, 1923, Nonmilitary Activities, Hearing before Subcommittee of House Committee on Appropriations, Pt. 2, Sixty-Seventh Cong., 2nd Sess.* (Washington: U.S. Government Printing Office, 1922), 184–85; Thomas Ewing Dabney, "Fighting the Water Hyacinth: Clearing the Clogged Waterways of Louisiana and Florida by Means of Live Steam," *Scientific American*, October 8, 1921, 260; "Massive Effort to Clear Bayou Here," *St. Mary and Franklin Banner-Tribune*, October 1, 2014, 1.

14. Mary Tutwiler, "Bayou Teche Fundraiser at Cafe Des Amis This Thursday," theInd.com, October 26, 2009, www.theind.com/article-4867-Bayou-Teche-fundraiser-at-Cafe-des-Amis-this-Thursday.html, accessed January 17, 2015; *Bayou Teche Scenic Byway, Corridor Management Plan, Vol. 1: Intrinsic Quality Inventory* (Lafayette, LA: Evangeline Economic Development District, June 2000), 5–7; Burke, *Crusader's Cross*, 57.

15. Dr. Beverly W. Smith et al., "The Investigation of Bayou Teche," *Biennial Report of the Louisiana State Board of Health to the General Assembly of the State of Louisiana, 1910–1911* (New Orleans: Brandao Printing, [1912?]), 139–40, 142, 144, 154, 198.

16. "Fish Kill Covers 20 Miles of Bayou," (New Orleans) *Times-Picayune*, October 23, 1975, Sec. 1, 11; "Bayou Teche Polluted by Mills, Residents Say," (New Orleans) *Times-Picayune/States-Item*, November 11, 1982, Sec. 6, 2.

DDT data is from Dennis K. Demcheck et al., *Water Quality in the Acadian-Pontchartrain Drainages: Louisiana and Mississippi, 1999–2001* (Reston, VA: U.S. Geological Survey, U.S. Department of the Interior, 2004), 19.

The latter source observed of a fish sample: "Concentrations of total DDT exceeded the NYSDEC criteria of 200 μg/kg [micrograms per kilogram] at Bayou Teche, reaching a maximum of about 350 μg/kg." 350 μg/kg is 75 percent more than the criteria of 200 μg/kg.

17. DDE data is from Demcheck et al., *Water Quality in the Acadian-Pontchartrain Drainages*, 18; and Stanley C. Skrobialowski, *Trace Elements and Organic Compounds in Bed Sediment from Selected Streams in Southern Louisiana, 1998, Water-Resources Investigations Report 02–4089* (Baton Rouge: U.S. Geological Survey, Department of the Interior, 2002), 19.

Carbofuran data is from *Louisiana Nonpoint Source Annual Report, Federal Fiscal Year (FFY) 2011* ([Baton Rouge?]: Department of Environmental Quality, [2011?]), 46.

Arsenic, chromium, and zinc data is from Skrobialowski, *Trace Elements and Organic Compounds*, 10 (Table 4).

Copper data is from "2002 Waterbody Report for Bayou Teche—Keystone Locks and Dam to Charenton Canal," Watershed Assessment, Tracking & Environmental ResultS (WATERS), U.S. Environmental Protection Agency, ofmpub.epa.gov/tmdl/attains_waterbody.control?p_list_id=LA-060401&p_cycle=&p_report_type=T, accessed January 8, 2015.

PAHs data is from Skrobialowski, *Trace Elements and Organic Compounds*, 19, 20 (Table 8).

Pathogen data is from "1998 Waterbody Report for Bayou Teche—Headwaters at Bayou Courtableau to I-10," Watershed Assessment, Tracking & Environmental ResultS (WATERS), U.S. Environmental Protection Agency, ofmpub.epa.gov/tmdl/attains_waterbody.control?p_list_id=LA-060205&p_cycle=&p_report_type=T, accessed January 8, 2015.

The Teche was referred to as an "impaired waterway" in "1998 Waterbody Report for Bayou Teche—Headwaters at Bayou Courtableau to I-10" and "2002 Waterbody Report for Bayou Teche—Keystone Locks and Dam to Charenton Canal," Watershed Assessment, Tracking & Environmental ResultS (WATERS), U.S. Environmental Protection Agency (see previous references in this endnote for complete citations).

The term "impaired waterway" is defined in numerous sources; see, for example, Whitney P. Broussard III, *Water Quality Monitoring in the Bayou Teche Watershed* (Lafayette: Institute for Coastal Ecology and Engineering, University of Louisiana at Lafayette, 2013), 1–2.

18. *Louisiana Nonpoint Source Annual Report*, 10, 46–47; Broussard, *Water Quality Monitoring in the Bayou Teche Watershed*, passim.

19. William Johnson, "Parish Council Rejects Naming Bayou Teche as Scenic River," theadvertiser.com, January 15, 2014, www.theadvertiser.com/story/news/local/2014/01/16/parish-council-rejects-naming-bayou-teche-as-scenic-river/4526239/, accessed January 17, 2015; William Johnson, "Landowners Oppose Special Designation for Bayou Teche," theadvertiser.com, February 7, 2014, www.theadvertiser.com/story/news/local/2014/02/05/landowners-oppose-special-designation-for-bayou-teche/5213321/, accessed January 17, 2015; Whitney Broussard, Lafayette, LA, e-mail to the author, January 14, 2015.

Teche Canoe Trip Journal

1. Rufus Jagneaux, "Port Barre," *Swamp Gold, Vol. 6* (compact disc, Jin Records JIN9064, 1999); Benny Graeff, songwriter, Sunset, LA, e-mail to the author, November 26, 2012. Lyrics used with permission of Flat Town Music Company, Ville Platte, LA.

2. Barthélémy Lafon, *Carte générale du territoire d'Orléans comprenant aussi la Floride Occidentale et une portion du territoire du Mississipi* (New Orleans: Barthélémy Lafon, 1806), original in the Geography and Map Division, Library of Congress, Washington; *American State Papers: Public Lands, Vol. IV* (Washington: Duff Green, 1834), 361–62; William Darby, *The Emigrant's Guide to the Western and Southwestern States and Territories* (New York: Kirk & Mercein, 1818), 51, 69.

Lafon's map can be viewed on the Library of Congress American Memories website at www.memory.loc.gov/ammem/index.html.

3. Prohibition-era bootlegging in the Leonville-Arnaudville-Pecanière area is mentioned in Barry Jean Ancelet, *Cajun and Creole Folktales: The French Oral Tradition of South Louisiana* (Jackson: University Press of Mississippi, 1994), lxvii; and David R. Williams III, *Gulf Coast Booze during Prohibition: Smuggling and Distilling*, Historical Text Archive, 1995 and 2010, www.historicaltextarchive.com/books.php?action=nextchapter&bid=72&cid=4, accessed July 3, 2015.

4. Glenn R. Conrad and Ray F. Lucas, *White Gold: A Brief History of the Louisiana Sugar Industry, 1795–1995* (Lafayette: Center for Louisiana Studies, University of Southwestern Louisiana, 1995), 14–15.

5. Perambulator [pseudonym], "Attakapas," *Louisiana Planter and Sugar Manufacturer*, October 11, 1890, 276; "Company Description," Levert Companies, n.d., www.levert.net/description.html, accessed July 5, 2015.

6. "Overview," Louisiana State Parks Historic Sites: Longfellow-Evangeline State Historic Site, n.d., www.crt.state.la.us/louisiana-state-parks/historic-sites/longfellow-evangeline-state-historic-site/, accessed July 5, 2015.

7. Herodotus, *The Histories*, trans. Aubrey de Sélincourt (New York: Penguin, 1986), 155.

8. "Mr. John Peters Disclaims Lowering the Level of Spanish Lake," *New Iberia Enterprise*, November 18, 1899, 2.

9. Darby, *The Emigrant's Guide*, 48, 58, 72.

10. J. B. Borrel, "The Verret Family de la Belle Place," *Attakapas Gazette* XXXI (Yearbook 1996): 67.

11. Smitty Landry, New Iberia, LA, e-mail to the author, October 6, 2014; Smitty Landry, *Write on the Teche: Memoirs of Growing up in New Iberia* (New Iberia, LA: Iberia Parish Public Library, 2006), DVD digital recording made October 19, 2006, St. Peter Street Branch, Iberia Parish Library, New Iberia, LA.

12. Alicia Duplessis, "Treasure—Old Ship—Found in Bayou," (New Iberia, LA) *Daily Iberian*, 13 January 2007, www.iberianet.com/articles/2007/01/13/news/news/news58.prt, accessed November 13, 2009.

13. Landry, *Write on the Teche*.

14. "Paper from Rice Straw Opens New Field," *Rice Journal* (January 1922): 28; Edmonds, *Yankee Autumn in Acadiana*, 44.

15. "Penal Hope," (Abbeville, LA) *Meridional*, December 29, 1900, 3; "Hope Plantation Chosen," *Donaldsonville Chief*, September 26, 1914, 1; Jennifer Koenig, Adam C. Viator, and Derik Wood, "Iberia Research Station Digital Archive Project Proposal," Louisiana State University School of Library and Information Sciences, ca. Spring 2011, classes.slis.lsu.edu/wu/7410/sp11/jkoenig/project_proposal.pdf, accessed November 10, 2015, 1, 4–5; Alcée Fortier, *Louisiana, Vol. II* (N.p.: Century Historical Association, 1914), 303; Nelwyn Hebert and Warren A. Perrin, *Iberia Parish* (Charleston, SC: Arcadia, 2012), 45.

16. Francis DuBose Richardson, "The Teche Country Fifty Years Ago," *Southern Bivouac* (January 1886); reprint, Glenn R. Conrad, ed., *Attakapas Gazette* VI (December 1971): 127.

17. "Tribal History," Sovereign Nation of the Chitimacha, n.d., www.chitimacha.gov/history-culture/tribal-history, accessed July 11, 2015.

18. Patrick Flanagan, "Kramer Goes from Gynecology to Genealogy," (New Iberia, LA) *Daily Iberian*, March 7, 2010, www.iberianet.com/news/kramer-goes-from-gynecology-to-genealogy/article_1cdbef83-3030-545d-9254-90fb6a14d725.html, accessed July 5, 2015; advertisement for lightning rods per the Charenton Post Office at Indian Bend, (Franklin, LA) *Planters' Banner*, August 30, 1849, 1.

19. Thomas Hutchins, *An Historical Narrative and Topographical Description of Louisiana and West-Florida* (Philadelphia: Robert Aitken, 1784; repr., Gainesville: University of Florida Press, 1968), 46; Dennis Gibson, ed. and anno., "The Journal of John Landreth," Pt. II, *Attakapas Gazette* XV (Summer 1980): 76.

20. "Surviving Steam Locomotives," Steamlocomotive.com, www.steamlocomotive.com, accessed July 26, 2015.

21. C. Ray Brassieur, University of Louisiana at Lafayette, e-mail to the author, March 18, 2012.

22. J. P. Hebert, Franklin, LA, e-mail to the author, July 21, 2014.

23. Routh Trowbridge Wilby, *Clearing Bayou Teche after the Civil War: The Kingsbury Project, 1870–1871* (Lafayette: Center for Louisiana Studies, University of Southwestern Louisiana, 1991), 88, 90, 91.

24. "Hanson Canal & Lock," American Canal Society Canal Index, October 6, 1981 [scanned typewritten document], American Canal Society, www.americancanals.org/Data_Sheets/Louisiana/Hanson%20Canal%20and%20Lock.pdf, accessed July 5, 2015.

25. Verdunville is identified as a Creole of Color enclave in Carl A. Brasseaux, "Creoles of Color in Louisiana's Bayou Country, 1766–1877," in James H. Dormon, ed., *Creoles of Color of the Gulf South* (Knoxville: University of Tennessee Press, 1996), 83–84.

26. Daniel Thompson obituary, *Chicago Historical Society, Report of Annual Meeting, November 20, 1900* ([Chicago: Chicago Historical Society, ca. 1900]), 315; Jane Adler, "New-Home Companion Being Forged for Old Prairie Avenue," *Chicago Tribune*, January 8, 1995, articles.chicagotribune.com/1995-01-08/business/9501080107_1_central-station-prairie-avenue-neighborhood/2, accessed July 6, 2015; A. T. Andreas, *History of Chicago, Vol. II* (Chicago: A. T. Andreas, 1885), 375.

27. "Less Danger, More Money!: New Rules, Classes for Tour du Teche," techetoday.com, August 4, 2010, www.techetoday.com/less-danger-more-money-new-rules-classes-tour-du-teche, accessed July 5, 2015.

28. Henry Rightor, ed., *Standard History of New Orleans, Louisiana* (Chicago: Lewis Publishing, 1900), 708.

29. R. Christopher Goodwin et al., *Historical and Archaeological Investigation of Fort Bisland and Lower Bayou Teche, St. Mary Parish, Louisiana* (New Orleans: R. Christopher Goodwin & Associates, June 1991), 143, 145.

30. Mark A. Rees, "Plaquemine Mounds of the Western Atchafalaya Basin," *Plaquemine Archaeology*, Mark A. Rees and Patrick C. Livingood, eds. (Tuscaloosa: University of Alabama Press, 2007), 84–87.

31. Jody Carinhas, Morgan City, LA, telephone interview by the author, July 27, 2013; "Patterson Menhaden Corporation, d/b/a Gallant Man, and Fletcher Miller, Agent; Surprise, Inc., d/b/a Surprise, and Fletcher Miller, Agent and Fishermen's Union Local 300, Amalgamated Meat Cutters & Butcher Workmen of North America, AFL-CIO," Case No. 15-CA-9475, September 29, 1965, in *Decisions and Orders of the National Labor Relations Board, Vol. 154, July 24 through September 30, 1965* (Washington: U.S. Government Printing Office, 1966), 1804.

32. Donald S. Frazier, *Thunder across the Swamp: The Fight for the Lower Mississippi, February 1863–May 1863* (Buffalo Gap, TX: State House Press, 2011), 132–40.

33. "The Sugar Business in Attakapas," *Report of the Commissioner of Patents, Agreeably to Law, Exhibiting the Operations of the Patent Office during the Year Ending December 31, 1845, 25th Congress, 1st Sess., Document 307*, in *Public Documents Printed by Order of the Senate of the United States, First Session of the Twenty-Ninth Congress, Vol. VI* (Washington: Ritchie & Heiss, 1846), 298; "Estate of Elam Patterson, Deceased," (Franklin, LA) *Planters' Banner*, December 6, 1851, 3; "Judicial Sale of Property Belonging to the Estate of Elam Patterson,

Deceased," (Franklin, LA) *Planters' Banner*, September 4, 1852; Carl A. Brasseaux and Keith P. Fontenot, *Steamboats on Louisiana's Bayous: A History and Directory* (Baton Rouge: Louisiana State University Press, 2004), 229.

34. Fulton C. "Butch" Felterman, recorded interview by author, Patterson, LA, October 25, 2013; Fulton C. "Butch" Felterman, Patterson, LA, e-mail to the author, August 7, 2015.

35. William T. Shinn, "The Rentrops and Their Ferry," *Attakapas Gazette* XXIII (Summer 1988): 59.

The 1811 legislation granting a ferry monopoly to Rentrop actually referred to "Lake de Jone" [*sic*]—but I believe this was a misspelling and that "de Jonc" was meant. "Jonc" is a Louisiana French word meaning "reed" or "bulrush" and was often used when describing south Louisiana's wetlands. It may be that Lake de Jonc corresponded to what is today called Grassy Lake, which sits between Lake Palourde and Lake Verret.

Index

Page numbers in *italics* indicate an illustration.

Unless otherwise stated all geographic locations are located in Louisiana.

About the Author

A native Louisianian and Cajun, Shane K. Bernard grew up in Lafayette and has lived for many years in New Iberia. He holds degrees in English and History from the University of Louisiana at Lafayette (where he currently serves as a Fellow of the Center for Louisiana Studies) and a doctorate in History from Texas A&M University.

Bernard is the author of four previous books about south Louisiana history and culture: *Swamp Pop: Cajun and Creole Rhythm and Blues* (1996); *The Cajuns: Americanization of a People* (2003); *Tabasco: An Illustrated History* (2007); and *Cajuns and Their Acadian Ancestors: A Young Reader's History* (2008).

For nearly a quarter century Bernard has served as historian and curator to McIlhenny Company—maker of TABASCO® brand products since 1868—and Avery Island Inc., both of Avery Island, Louisiana.

He has been interviewed about south Louisiana history and culture by numerous media outlets, including the BBC, CNN, NPR, The History Channel, the *New York Times*, and *National Geographic*.

Bernard is married with two children and resides a short distance from Bayou Teche.